SpringerBriefs in Electrical and Computer Engineering

Series Editors

Woon-Seng Gan, School of Electrical and Electronic Engineering, Nanyang Technological University, Singapore, Singapore

C.-C. Jay Kuo, University of Southern California, Los Angeles, CA, USA

Thomas Fang Zheng, Research Institute of Information Technology, Tsinghua University, Beijing, China

Mauro Barni, Department of Information Engineering and Mathematics, University of Siena, Siena, Italy

SpringerBriefs present concise summaries of cutting-edge research and practical applications across a wide spectrum of fields. Featuring compact volumes of 50 to 125 pages, the series covers a range of content from professional to academic. Typical topics might include: timely report of state-of-the art analytical techniques, a bridge between new research results, as published in journal articles, and a contextual literature review, a snapshot of a hot or emerging topic, an in-depth case study or clinical example and a presentation of core concepts that students must understand in order to make independent contributions.

More information about this series at https://link.springer.com/bookseries/10059

Navneet Ghedia • Chandresh Vithalani
Ashish M. Kothari • Rohit M. Thanki

Moving Objects Detection Using Machine Learning

Navneet Ghedia
Sanjaybhai Rajguru College
of Engineering
Rajkot, Gujarat, India

Ashish M. Kothari
Atmiya University
Rajkot, Gujarat, India

Chandresh Vithalani
Government Engineering College
Rajkot, Gujarat, India

Rohit M. Thanki (iD)
Prognica Labs
Dubai, United Arab Emirates

ISSN 2191-8112   ISSN 2191-8120   (electronic)
SpringerBriefs in Electrical and Computer Engineering
ISBN 978-3-030-90909-3   ISBN 978-3-030-90910-9   (eBook)
https://doi.org/10.1007/978-3-030-90910-9

This Springer imprint is published by the registered company Springer Nature Switzerland AG
The registered company address is: Gewerbestrasse 11, 6330 Cham, Switzerland

Preface

In the area of computer vision, ensuring high-level security in public places is extremely important and challenging. Providing a high level of vigilant traffic monitoring, vehicle navigation, and autonomous driving assistance is also a crucial task in the era of the computer vision and surveillance systems. In surveillance applications, object detection and tracking are significant and essential tasks. Researchers have chosen visual surveillance because of its importance in military applications, law enforcement, robotic perception and navigation, and crowd analysis in the last few years. We present to use a fixed pan-tilt-zoom (PTZ) camera in public places to provide a high level of security, as well as developing a monocular and 2D object and 3D object detection and tracking algorithm.

In this book, we discuss the different background subtraction approaches, foreground segmentation, and object tracking approaches. The presented algorithm addresses a multimodal background subtraction approach that can handle dynamic backgrounds and different constraints. Using the algorithm presented, it is possible to detect and track both 2D and 3D objects in monocular sequences in both indoor and outdoor surveillance environments. At the same time, it can work satisfactorily with dynamic backgrounds and challenging conditions. In addition, the present algorithm uses parameter optimization and adaptive thresholds as intrinsic improvements over the Gaussian Mixture Model.

Overview of the Book

In Chap. 1, basic information about video surveillance systems is presented. Chapter 2 addresses a brief literature survey on background model estimation, various parameter initialization algorithms, motion segmentation techniques, and a detailed examination of object tracking methodologies. In Chap. 3, an indoor and outdoor probability model is presented, along with the necessary derivations. The chapter presents algorithms for parameter initialization and maintenance.

Furthermore, it provides 3D monocular object detection and adaptive thresholding for foreground detection. Various approaches for tracking objects are provided in Chap. 4. The chapter also discusses the Kalman filter's mathematical expression and its mathematical description.

Features of the Book

- Basic information of video surveillance systems
- Various approaches for background modeling and foreground modeling
- Detail for detection of 2D object and 3D object in video frame
- Detail of Kalman filter along with its application for object tracking in video frame

Acknowledgments

The authors would like to express sincere gratitude to Mary James, Senior Editor, Springer, for continuous guidance, motivation, and encouragement and for providing the wonderful opportunity to publish our work. We would also like to thank the production department of Springer for providing many valuable and constructive comments to effectively finish the last mile.

Rajkot, Gujarat, India   Navneet Ghedia
                                                                                    Chandresh Vithalani
                                                                                     Ashish M. Kothari
Dubai, UAE   Rohit M. Thanki

Contents

Chapter 1
Introduction

1.1 Introduction

Visual monitoring system is the most researched topic in today's era. Visual observation in computer vision helps to analyze the object's activities easily. Today's era of computer vision will completely remove traditional human-operated video surveillance system. A major part of smart video surveillance system is characterized by perception; the robustness of a smart video surveillance system (SVSS) is not only to sense the environment but also to interpret and act intelligently. Advancement in perception will lead to applications for defense and automated driving assistance. Nowadays researchers are working on object detection, object tracking, crowd analysis, pedestrian, and vehicle identification to improve the security at the public places.

T. Reeve [1] and Rajiv Shah [2] have extensively surveyed the penetration and importance of the surveillance system in United Kingdom (UK) and United States (USA), respectively. They have reviewed that the large amount of surveillance data monitoring was done by human operators over a longer time and yet it does not yield vigilant monitoring. Modern researchers are putting more concentration on real-time processing of visual observations because of the tremendous growth of computers and low-cost high-resolution cameras. Over more than a decade, researchers focus their attention not only to object detection and object tracking for smart surveillance system but also on a real-time processing multiple cameras, and even on more recent developments, they focus on 3D object detection and tracking. For 3D object detection and tracking, people are using different approaches for depth calculation. Some of the methods are using stereovision for depth calculation using the disparity, while some are using monocular clues for 3D detection. There is a clear tradeoff among two approaches of computational v/s depth precision. Some of the current methods completely estimate 6D pose of the object.

© The Author(s), under exclusive license to Springer Nature Switzerland AG 2022
N. Ghedia et al., *Moving Objects Detection Using Machine Learning*, SpringerBriefs
in Electrical and Computer Engineering,
https://doi.org/10.1007/978-3-030-90910-9_1

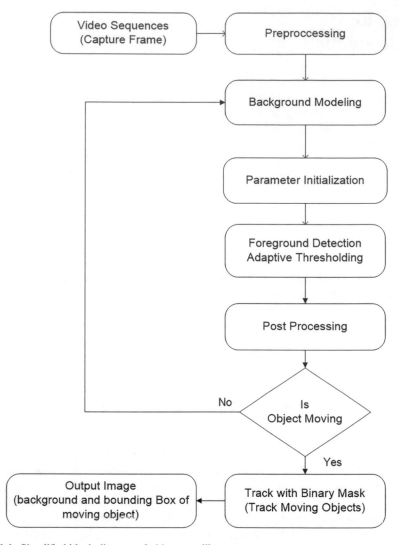

Fig. 1.1 Simplified block diagram of video surveillance system

Figure 1.1 shows the primitive operations of the video surveillance system. The automated surveillance system requires different components. The incoming video sequences may be with or without the background, so it is necessary to generate background with the help of the subsequent frames or by means of frame analysis. Primitive operation preprocessing is required to remove the dataset noises and outliers. One of the important components of every surveillance system is to establish the background model which is appropriate for every video sequence. The robustness of background modeling gives accurate foreground detection. Finally, tracking can be achieved with the help of the tracker.

Fig. 1.2 Example of an indoor surveillance system

Fig. 1.3 Example of an outdoor surveillance system

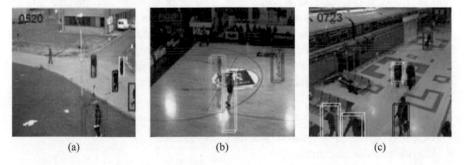

Fig. 1.4 Example of monocular 3D tracking

A typical visual surveillance system consists of static pan-tilt-zoom (PTZ) video camera, and it will transmit video sequences to a central surveillance room or is stored in a surveillance monitoring server. Such video sequences are being observed carefully by human operators. If monotonous monitoring activity of an operator might miss some important incidences at that time, then such loss in lapses during monitoring activities becomes challenging. Examples of visual surveillance systems are shown in Figs. 1.2 and 1.3, while Fig. 1.4 shows monocular 3D tracking. Any smart surveillance system must be able to perceive and identify the new scene because a human operator is not able to change the algorithm parameters every time. The system can operate in different environmental conditions like sudden changes in illumination, clutter background, occlusions, etc. with minimum error. Such system is called a robust video surveillance system.

A single camera surveillance system suffered from many issues like occlusions, different silhouettes of still and moving object shadows, weather effects, dynamic background scene, etc. Though the single camera endured all such constraints, still it is preferable to some extent because of its limited capability and its affordable price. To handle these constraints, the use of multicamera system is preferable instead of single camera for surveillance system. Even 3D detection and tracking can be achieved by means of single or multiview monocular video sequences. The depth maps can be generated by means of stereoscopic or monocular clues.

1.2   Challenges in Video Surveillance System

Situation alertness is key to security. There are basically three kinds of work that a security analyst needs to "see, check, and track." They must identify the people and vehicle in space and locate the people and what kind of activity they do in space. They also use chronological context to understand the statistics taken from the above knowledge. Enhancing and ensuring a fair level of security across multiple scales of time and space in public places such as airports and railway stations and at other places becomes an extremely difficult challenge for smart video surveillance system and it also increases situational awareness. There are multiple security challenges like screening system, database system, biometric system, and video surveillance system for object tracking and verifying identity and monitoring activities, respectively. Today's video surveillance system focusing on compression for the purpose of storing and transmitting performs as either an analog or digital video recorder. Locating, identifying, and learning the object behavior in video sequence require three main steps:

- Detecting foreground objects that are in motion
- Detected objects to be tracked in consecutive frames
- Object behavior recognition

Smart video surveillance system requires fast and robust algorithm for estimating background, motion segmentation, object tracking, and scene analysis, and it also assists operator for important scene events. Smart and intelligent video surveillance is the most researched topic for the last decade because more importance is given to security and military applications [3].

Moving object detection is the primitive operation for scene analysis. For accurate object tracking, one must be able to detect foreground objects precisely in every frame under different conditions. Objects should be detected as soon as it appears in the frame. In some cases where the exact background cannot be achieved in the earlier stages, the algorithm requires additional frames to create background or

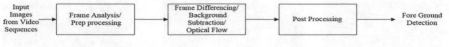

Fig. 1.5 Simple background estimation approach

Fig. 1.6 Simple object tracking approach

extract the background information under nonstationary background, and in such cases, we may lose the tracking accuracy for those frames. The robustness of every tracking algorithm depends on the successful detection of foreground. Some of the popular object detection techniques are background subtraction, frame difference, and optical flow.

Frame difference algorithm detects objects by evaluating consecutive frames; it gives less computational time, but it cannot adopt dynamic background. Background subtraction can be implemented by simply subtracting out background frames from every frame, but it is not able to handle dynamic backgrounds. Optical flow is an alternative approach to deal with different background constraints, but such a technique requires more computational time as compared to that of other approaches. In some surveillance situations both the camera and foreground objects are moving; optical flow is the best approach to handle such situations and to handle the dynamicity of the background. Instead of the traditional background estimation, mixture model gives more robustness and can also handle different backgrounds. Figure 1.5 shows the block diagram of simple background estimation approach.

Robust and accurate tracking of nonrigid, complex, fast-moving objects is a major challenge nowadays. Object tracking is a vital task in surveillance system. It is also applied in video editing, augmented reality, traffic monitoring and control, gesture, posture recognition, etc. The challenging task is to track object in different conditions like illumination variations, dynamic background, complex object silhouette, and occlusions. A robust tracking must predict the positions of the object whether it is being occluded or not and ensure its position, making sure that it will not lose the object completely. Multiple views and cameras can handle such conditions. One of the major motivations for tracking is the ball tracking system 'Hawk-Eye' [4]. It uses object tracking techniques to track the ball in cricket as well as in tennis sport [5] and proposed two different tracking approaches named top-down and bottom-up. The bottom-up approach is generally focused on typical application like data mining, gesture recognition, and sports events. While the top-down approach is generally used for surveillance purposes. In first-world countries, smart and real-time surveillance systems used the top-down approach. The top-down approach consists of foreground segmentation, motion detection, and object tracking (Fig. 1.6).

      (a)   (b)   (c)

Fig. 1.7 Low frame rate video sequences (consecutive frames)

Figure 1.7 shows low frame rate video sequences. It would be very challenging for every video surveillance system to segment the moving foregrounds and track the moving objects. It indicates that video sequencing is a demanding task and gives an ample amount of motivation to every researcher. The racking algorithm should also be robust in terms of not only handling the constraints but also operating on both indoor and outdoor environments. Template matching, mean shift, motion estimation, Kalman filtering, particle filtering, and silhouette tracking are some of the popular tracking approaches available for object tracking.

1.3   Contributions in This Book

This book aims to present a novel background modeling technique for foreground detection in both the indoor and outdoor environments. The book gives information regarding the development of object detection algorithm which can detect and track 2D and monocular 3D information in visual surveillance system using probabilistic statistical approach. Also, to ensure high level of security in public places using static PTZ camera, robust detection and tracking algorithm for video sequences easily adopt background changes. In addition, the book gives performance analysis of the detection and tracking system under different challenges. The presented algorithm provides possible intrinsic and extrinsic improvements to the challenges in the video surveillance system as follows:

1. Intrinsic Improvement: Motion Segmentation

   - Select model parameters appropriately using parameter optimization algorithm.
   - Develop a robust foreground detection algorithm through adaptive thresholding for motion segmentation.

2. Extrinsic Improvements: Performance Evaluation

- Improve performance evaluation parameter using postprocessing technique.
- Remove dataset noise by preprocessing technique.

Moreover, the book also provides various information on research work which targets the below constraints:

- Performance under different backgrounds
- Handling varying lighting conditions such as bright, dark, low, and high contrast
- Invariance to camera perspective
- Handling process and data noise
- Working on various resolutions and video sequences with different frame rates
- Handling partial occlusion and detecting and tracking near-field, mid-field, and far-field objects
- Detecting and tracking objects with similar appearance, with different height, and with different motion
- Detecting and tracking 3D multiple objects in monocular video sequences
- Handling crowded scenes

The book gives a unifying algorithm for 2D and monocular 3D object detection and tracking. The presented algorithm integrates different modules like modified Gaussian mixture model, adaptive thresholding for motion segmentation, and Kalman filtering for both indoor and outdoor surveillance systems. The presented algorithm is unique and simple in reference to other state-of-the art algorithms. It is capable of handling different background dynamics and constraints. Robustness in terms of handling the different constraints is the major challenge and it is achieved by the presented algorithm. This book provides research in following areas:

- *Intrinsic Improvement (Parameter Optimization Algorithm)*: Intrinsic improvements in Gaussian mixture model which concern on the modification made in Gaussian model parameter initialization and parameter maintenance during execution at every new pixel or frame level and at the foreground detection (motion segmentation) level. The appropriate selection of mixture parameter is indeed an impact on the performance of the overall surveillance system, as the same algorithm is applicable for both indoor and outdoor surveillance systems. The existing work shows that usually, the model parameters are predefined or initialized by some algorithms such as k-means cluster algorithm, EM or MLE algorithm, etc. In the presented algorithm, model parameters can be initialized by parameter optimization algorithm for every video sequence. The presented algorithm is evaluated with the standard video datasets and compared with the other existing algorithms and significant improvements are obtained as outcome.
- *Foreground Detection (Motion Segmentation)*: Foreground detection plays a vital role in surveillance system. The background model is sensitive enough to segment every moving object. The existing work shows different intensity, region, texture, edge, or motion-based segmentation algorithms. Static threshold provides poor foreground detection and may lead to increases either false positive or false negative. In the presented algorithm foreground detection is achieved by means of adaptive thresholding instead of static thresholding. The presented

algorithm is evaluated with the standard dataset and the resultant foreground mask is compared with the ground truth and other similar approaches. The observation is that most of the false negatives generated by the traditional pixel- and region-based methods are removed by intrinsic improvements.

- *Extrinsic Improvement (Preprocessing):* Extrinsic improvements emphasize purely on improving the performance of the model and hence it also tends to improve the results. The image and dataset noises are removed by using the preprocessing. The presented algorithm used adaptive local noise reduction filter as a preprocessing method to remove dataset noises.
- *Extrinsic Improvement (Postprocessing)*: Postprocessing is again an external tool to perform the evaluation. In the presented algorithm, morphological closing (dilation followed by erosion) is being used as a postprocessing method for the sake of reducing the noise and outliers in the datasets. As a result, most of the false positives generated by traditional approaches are removed.

For performance analysis of the presented algorithm, this algorithm shows significant improvements in well-known and widely used parameters such as similarity measures, object detection, tracking accuracy, recall, and precision as compared to other similar pixel-based parametric, nonparametric, and region-based background subtraction counterparts.

1.4   Book Organization

The problem in indoor and outdoor surveillance is described in the subsequent chapters and that can detect and track objects in both indoor and outdoor environments and manage dynamic backgrounds and partial occlusion. The chapters are structured as follows:

- Chapter 2 *Survey on Video Surveillance System*: This chapter provides a brief and general literature survey on background model estimation, various parameter initialization algorithms, various background modeling methods, motion segmentation approaches, and a detailed investigation of object tracking approaches.
- Chapter 3 *Background Modeling:* This chapter provides background model and necessary derivations related to probabilistic model for both indoor and outdoor environments. The chapter provides parameter initialization algorithm and maintenance techniques. Also, it provides foreground detection using adaptive thresholding and 3D monocular object detection.
- Chapter 4 *Object Tracking:* This chapter provides the concept of object tracking and its various approaches. The mathematical expression and discussion of the Kalman filter are also covered in this chapter.
- Chapter 5 *Summary of the Book:* This chapter provides a summary of the presented approach with comparative results and provides certain suggestions on supplementary investigation on video surveillance system.

References

1. Reeve, T. (2011). How many cameras in the UK? Only 1.85 million, claims ACPO lead on CCTV. *Security News Desk*. Weblink: https://securitynewsdesk.com/how-many-cctv-cameras-in-the-uk/. Last Access: July 2021.
2. Shah, R. (2011). Effectiveness of redlight cameras in Chicago. Web archive: https://thenewspaper.com/rlc/docs/2010/il-chicagostudy.pdf.
3. Collins, R. T. (2000). *A system for visual surveillance and monitoring*. Final Report CMU-RI-TR-00-12.
4. Owens, N. E. I. L., Harris, C., & Stennett, C. (2003, July). Hawk-eye tennis system. In *2003 international conference on visual information engineering VIE 2003* (pp. 182–185).
5. Zhang, H. J. (2004). New schemes for video content representation and their applications tutorial. In *IEEE international conference on image processing*.

Chapter 2
Existing Research in Video Surveillance System

2.1 Introduction

In an ideal video surveillance system, background is expected to be static in every frame and foreground or moving object is expected to be moving in each frame. In real-time analysis and monitoring system, background is changing over time, generally for outdoor scenes. Sometimes it may also become stationary for longer duration, i.e., for cars and persons. In recent times, monitoring foreground objects such as vehicle and people has a wide application in vehicle navigation, traffic monitoring, automated driving assistance, and crowd analysis and interactions among strangers and so on. Literature shows the general outline and different component and approaches of the surveillance system for object detection and tracking achieved under different dynamic environments. Video surveillance and monitoring (VSAM) [1] is a US government-funded system used to detect and track moving vehicles and pedestrian. It also detects activities and interactions among objects. The KNIGHT system is proposed by Shah et al. [2]. It is a real-time automatic surveillance for object detection, classification, and tracking and analyzes object behaviors using single and multiple camera systems. Y. Tian et al. [3] proposed the IBM smart surveillance system (S3) to detect and track moving objects especially in monitored area. Some of commercially available surveillance systems are VistaScape (Siemens), Acuity, Avocado, AxonX, Axis, ioimage, Mate, etc. for object identification and for monitoring activities carried out in a space. Sections 2.2, 2.3, 2.4, and 2.5 include reviews of methods for 2D objects and Sect. 2.6 includes that for 3D objects.

2.2   Gaussian Mixture Model-Based Background Modeling and Its Improvements

In this section, various Gaussian mixture model-based background modeling approaches available in the literature are discussed. Stauffer et al. [4] proposed a popular background modeling Gaussian mixture model (GMM) approach. They have suggested a multivariate background model using GMM for each pixel. They compared each pixel of frames to the GMM for determining the foreground and background component. This approach gives robust detection against illumination variations, clutter background, and moving background such as breeze in trees, water twinkling, etc. It however fails against sudden changes in illumination. The traditional GMM approach has taken more computational time. Bowden et al. [5] proposed a system that overcomes the demerits of that in Stauffer et al. [4] such as slow learning rate. The authors have explained that improvements in learning rate will converge faster in a stable background. They have also proposed a system to deal with shadows. Wang et al. [6] suggested an additional approach for shadow removal and background updation. Harville [7] proposed a system which performed foreground segmentation based on depth and color features. They have used traditional mixture of Gaussian (MoG) approach for foreground detection. Again, the algorithm suffered from slower learning rate so it is not able to respond immediately to illumination change.

One of the major changes in surveillance system is the capacity to clearly identify the foreground and background in every frame. All the previous approaches incurred learning rate issues. In a video sequence foreground or moving objects are not part of the background. However, if foreground objects stop still, then it becomes identified as a foreground object, and if the same foreground object becomes still for a longer duration, then it will be incorporated into the background. The Gaussian mixture model is updated at every new pixel/frame with the help of learning parameters. The slow learning rate will not destroy the existing background model, but it will update the background model and incorporates still moving objects. The challenge in slow learning rate is that it will not be able to handle sudden changes in light conditions. Hence, for outdoor surveillance where light condition becomes the highest constraint, such an approach would not be applicable.

To overcome such problem, Harville [8] has proposed an approach which is an extension of Gaussian mixture model. The author allows certain high-level feedback to adopt the fast changes in background. For almost all stationary objects, the weight of the mixture remains low so it should be updated, and the object can still be correctly identified as a foreground. Wren et al. [9] proposed Gaussian background modeling. Every background pixel of a scene can be estimated as a Gaussian distribution, where background color and covariance matrix can be defined as the mean and covariance. Mahalanobis distance can be used to identify the distance metric. Cheng et al. [10] proposed the probabilistic mixture of Gaussian (MoG) background model. They have recursively estimated the MoG parameters and obtained a number of Gaussians. They have proposed expectation-maximization

(EM) algorithm for parameter initialization and for maintenance. The proposed algorithm can detect foreground object under critical light conditions.

Shimada et al. [11] proposed the MoG approach to handle dynamic environments and the number of Gaussian changes automatically for each pixel. As pixel values change, the number of Gaussian increases, and when pixel values are constant, Gaussians are eliminated or incorporated. The MoG parameters can be maintained by IIR filters. Tan et al. [12] proposed a GMM algorithm in which the number of Gaussian can be estimated at each pixel. The model parameters are initialized by online EM algorithm and maintained by IIR filtering to make it adaptive. Carminati et al. [13] proposed a GMM algorithm in which the number of Gaussians can be optimally estimated using ISODATA algorithm. The proposed approach gives limited adaption as it is restricted by the training period. Pavlidis et al. [14] proposed a unique Gaussian mixture model approach to adopt the EM algorithm for initialization. Such an algorithm provides faster learning and stability to GMM algorithm for accurate foreground detection. This approach is computationally costlier than others. Lee [15] proposed an approximate EM algorithm for moving object detection. It can avoid unnecessary storage and computations.

Zhang et al. [16] proposed a novel approach for GMM parameter initialization. They have suggested background reconstruction algorithm to initialize MoG in the presence of moving objects in the frame. Aminatoosi et al. [17] proposed a robust GMM algorithm by adopting QR decomposing algorithm. Such algorithm robustly handles occlusions. Morellaset et al. [14] recommended intrinsic improvements in MoG. They have suggested an algorithm for model parameter initializations. The proposed approach handles dynamic and clutter background. White and Shah [18] explained the better utilization and importance of learning rate for parameter maintenance. Javed et al. [19] proposed a system to overcome the problem of slow learning rate. They have used gradient information of MoG to overcome illumination changes. In a sudden illumination change, the color of the background may change but it will not affect the gradient of the background. So, we can easily calculate various Gaussians that are present in gradient distributions of each pixel. If the computed gradient is matched with one of the distributions, then the pixel becomes the background pixel.

Zang et al. [20] proposed a GMM called pixel map. They have combined MoG with segmentation approach. They have proposed region- and frame-level approach for noise reduction and elimination of holes. Butler et al. [21] developed an algorithm for background segmentation. They have used a pixel which is modeled in a form of cluster and again it consists of color, centroid, and weight. This cluster-based adaptive background segmentation approach takes more computational cost. Zivkovic et al. [22] presented a novel approach of AGMM technique to adopt the scene repeated. In contrast to traditional GMM, the proposed algorithm adaptively chooses the number of distributions of GMM. It can handle multimodal background and dynamic environments. Lee [23] has briefly explained the importance of learning rate. The new adaptive learning rate is required to improve the convergence rate without affecting the stability of GMM.

Landabaso et al. [24] proposed an approach for the improvements made in foreground detection. Robust foreground detection can be achieved by using Pixel Persistence Map (PPM) instead of simple intensity-based approach. Schindler et al. [25] explained that the efficiency and robustness of the GMM algorithm can also be achieved by extrinsic improvements. They have suggested a well-known Markov random field (MRF) approach for postprocessing to improve efficiency.

2.3 Pixel-Based Background Modeling

Pixel-based background modeling depends on histogram statistics or on other probabilistic approaches. We can also easily determine the pixels belonging to foreground or background. Ma et al. [26] have proposed a robust crowd analysis approach. They have evaluated a line relationship between the foreground pixels and persons. For complete isolation among moving objects, foreground pixels become equal to the moving object. Chan et al. [27] proposed a traditional frame difference algorithm for moving object detection. Such an algorithm cannot handle nonstationary background, occlusion, and dynamic environment. Lipton et al. [28] suggested a novel pixel-wise frame difference approach. This algorithm can handle dynamic scene but fails to extract all foreground pixels. The binary mask generated by the algorithm suffered from holes. Collins et al. [1] proposed a novel hybrid frame difference approach to handle a dynamic scene. Instead of two frames, they used three frames differentiating the adaptive probabilistic background subtraction for their video surveillance and monitoring research work. Jacques et al. [29] proposed novelty in shadow detection by estimating background models in grayscale video sequence. Median filter is required to differentiate moving and static pixels. Static pixels are playing a vital role in estimating background and they are also responsible for removing the shadow from foreground pixels.

Horitaoglu et al. [30] proposed a W^4 system to adopt noise locally. The approach works on color model and uses frame difference min-max method. Kim et al. [31, 32] recommended a codebook method that uses a clustering/quantization technique to estimate multimodal backgrounds. The codebook approach keeps records of several code words for every pixel. It is a kind of nonparametric pixel model where every codeword is a series of key color values. The authors have proposed cache book like codeword used to reduce false negatives and positives as compared to other traditional parametric approaches. The codebook approach can handle or retain background motion for a longer period. The proposed approach can detect moving object in the presence of illumination change but it cannot handle moving objects with similar appearance.

Oliver et al. [33] proposed eigen background used to model each background pixel. It is a kind of pixel based on parametric modeling technique. An eigen background model provides robustness to the probability distribution function. The proposed algorithm robustly detects the small silhouette of foreground objects, and it also removes the outliers of foreground objects. The updation of the background model is computationally intrusive to perform. Xu et al. [34] proposed improvements in eigen background model. They suggested using recursive error compensation approach to diminish the influence of foreground object on the eigen background model for background subtraction. Another possible suggested improvement is to adopt adaptive threshold. The combined improvements give robustness to the algorithm in the presence of foreground objects (silhouette).

2.4   Region-Based Background Modeling

In region-based background modeling method, a pixel cannot be considered as a part of an object or its intensity values. Pixels are measured based on their connectivity with their neighbors of the same region. Elgammal et al. [35] introduced kernel density estimate on pixel intensity values for background estimation. The proposed approach is a nonparametric method because of the introduction of kernel function. It can handle multimodal backgrounds. It can also adopt fast illumination changes compared to that of other parametric approaches and the algorithm requires a large memory. Russell et al. [36] proposed a unique block matching region-based approach to discriminate foreground object from the background. The authors have compared the image region of new frame with the existing fixed size database of background. Such approach can be able to handle outdoor dynamic backgrounds.

Ma et al. [26] detect foreground objects by using random homogeneous region movements and preprocessing of image regions in CIELAB color space. Heikkila et al. [37] proposed a background model on the basis of discriminative texture feature called local binary pattern (LBP). They have developed LBP histogram for overlapping regions of background and compared them with LBP histograms of each region of incoming frames. Seki et al. [38] proposed region-based background models based on co-occurrence of image variations in background. Liu et al. [39] recommended a unique foreground detection approach which detects foreground under illumination variations using binary descriptor-based background model. The multimodality of the kernel density estimator (KDE) detects multimodal scenes/backgrounds especially in fast illumination changes (tree waving, water twinkling, etc.). The major drawback of KDE is that it requires N frame memory to train the background model.

Travakkoli et al. [40] and Tanaka et al. [41] have proposed probable improvements that could be made in KDE to deal with the time/memory constraints. They [40] suggested an intrinsic improvement such as changing the kernel function and proposed a selective background maintenance schedule to reduce computational time. They [41] also proposed an approach to decrease the sample by selecting/determining the proper size of the frame buffers. Mao et al. [42] recommended a selective diverse sampling background maintenance approach to reduce computational time. So, the computational time constraint of KDE algorithm can be managed by maintaining background using recursive or selective methods. Further improvement in region-based KDE algorithm is made possible by improving foreground detection approach. Ianasi et al. [43] have proposed improvements in background model by using constant kernel bandwidth, and they have recommended a unique dissimilarity measure for foreground detection. Some authors have suggested probable improvements in extrinsic model parameters to handle the constraints and proposed enhancement/improvements in foreground detection.

2.5 Hybrid Background Modeling

A hybrid method combines two different methods, i.e., pixel based and region based. Generally, pixel-based background modeling provides better representation and region-based approach handles dynamic background. So, the hybrid model can detect and handle dynamic scenes and illumination changes. Patwardhan et al. [44] proposed a novel approach for motion detection. They have used such approach to locate the layers within a scene. Generally, the same color values are accommodated in each layer. At the time of frame processing based on likelihood estimation, each frame can be compared to previous frames. An algorithm continuously compares the pixels with the layers. If the pixel does not match with the background layer, then it is considered as a part of foreground layer. It also maintains the existing layer and creates a new layer. Likewise, according to Stauffer et al. [4], if the moving object remains still for a longer duration, then the foreground layer is added to its background layer, and if the same still object starts moving once again, its background layer is removed and the foreground layer is updated.

Abdelkader et al. [45] have proposed a unique approach for robust motion detection by thresholding and background subtraction methods. It recursively modifies each frame with the mean and variance of pixel. The proposed approach doesn't require a training period to develop the background model. Grabner et al. [46] recommended a novel approach for background subtraction and they have also introduced the online AdaBoost method. The input sequences are divided into rectangles and a boosted classifier. Such a system achieves good results in low illumination scenes. The proposed algorithm performs offline detection and tracking. Toyama et al. [47] suggested background estimation by applying the linear predictive filters. K pixel samples will predict the respective background. Generally, Wiener filtering would be the suggested approach for such an environment. After a match test, if the predicted value of the pixel is beyond certain threshold value, then it can be treated as a foreground. Koller et al. [48] presented a unifying silhouette-based approach for vehicle tracking. The object silhouette is obtained from gradient image and an adaptive background model is used to locate vehicles on the road.

Marana et al. [49] explained how to represent crowd density levels. They have used coarse and fine textures to represent low and moderate density crowd. Wang et al. [50] proposed a pixel-based nonparametric statistical approach to estimate background. This approach keeps records of previous observations of each pixel and classification of new pixel significance as a background and calculates match points. The algorithm handles pixel- and blob-level strategies to deal with the entire object.

Maddalena et al. [51] proposed an artificial neural network based on self-organizing background subtraction (SOBS) method for multimodal background modeling. The proposed algorithm can also handle gradual illumination, dynamic background, and camouflage in various static video sequences. Barnich et al. [52] proposed a novel random strategy in background modeling to estimate background using sample-based estimation. This background model can be updated using non-recursive method. The proposed algorithm is a pixel-based nonparametric approach that uses color value of every pixel to estimate the background. This approach cannot adopt sudden illumination charges and requires manual adjustment of parameters to handle dynamic scenes.

Hofaman et al. [53] have explained a novel pixel-based adaptive segmenter (PBAS) approach to detect foreground objects. The background model is initialized by N image and updated randomly. The segmenter approach uses adaptive parameter instead of fixed parameter. The lower and upper segmenter boundary handles dynamic and non-static background. Huang et al. [54] suggested a novel approach of modeling by combining pixel-based RGB colors with optical flow motions. The hybrid model significantly segments foreground objects from static or non-static backgrounds. Such hybrid model gives computational complexity compared to that of other pixel- or region-based approaches. T. Sai et al. [55] presented a novel hardware approach for foreground detection: a hybrid algorithm. Chen et al. [56] proposed a hybrid model to improve the performance evaluation of the GMM algorithm by suggesting extrinsic improvements. They have recommended a hierarchical approach for postprocessing.

2.6   3D Object Detection

A large part of robot independence is characterized by awareness, the ability of a robot to sense its surroundings and act in an intelligent way. 3D scene understanding, object detection, and tracking are among the greatest challenges in computer vision. Structure from motion, optical flow, stereo, edge detection, and segmentation are some of the methodologies which are able to understand and recognize the object. From a monocular image, 3D reconstruction requires interpreting the pixels in the image as surfaces with depths and orientations. Generally, perfect depth estimates are not necessary; a rough sense of geometry combined with texture mapping provides a convincing detail.

Household robots and cars require path finding, detecting obstacles, and predicting other object movements within the scene space. Mobile robots and autonomous vehicles will require image analysis along with depth sensors. Image analyses require good abstractions for interaction and prediction in 3D coordinates. Much of the prior work on 3D object detection has focused on offline computation to prevent real-time tracking portions. Foreground object information can be defined and detected by two different approaches: (1) appearance-based model (2) geometry-based model.

- *Appearance-based model*: Appearance models are formed by volume-based silhouette technique. This model is trained based on the silhouettes of every object. (Features/descriptors are derived from the images of the object utilized.)
- *Geometry-based model*: Geometric models are constructed for every object and shape-matching techniques. Geometry-based methods handle occlusions and similarity appearance because it does not depend on the appearances. 3D object detection and tracking approaches have undergone remarkable improvements in recent years. Generally, all the approaches depend on depth sensors, feature points, texture, and edge.

Dementhon et al. [54] proposed POSIT algorithm to determine the pose of an object by matching the feature points to geometric point on object. This algorithm is implemented on monocular scenes, and it is fast to implement. Such an algorithm fails to handle mismatches. Najafi et al. [57] provided an extended version of the previous approach and it can handle mismatches. The proposed approach utilizes a set of calibrated images of moving objects with 3D geometric model. The feature correspondence is achieved by a combining effort of Bayesian framework and principal component analysis. Alberto et al. [58] proposed a real-time pose estimation algorithm able to handle challenging conditions. The algorithm is able to detect and track 3D object in monocular images. It does rely on depth sensor and depths can be estimated from the 2D projections. Xiang et al. [59] proposed an algorithm which can track 3D object in monocular multi-view scenes. The approach performs 3D tracking through part-based particle filtering and utilizes the monocular multiview tracking. Such an algorithm is not robust to occlusion.

Brachmann et al. [60] proposed 6D pose estimation approaches on the base of depth sensor. It utilizes the 3D coordinates to estimate 6D pose. This approach depends on the recognition of local patches. It is considered for RGB-D images. Chliveros et al. [61] proposed that contours and edges are used to estimate pose. Such an algorithm cannot handle occlusions. Tejani et al. [62] recommended an approach to estimate and detect 3D object and pose in monocular image. This algorithm uses local patches for object recognition, and it fails to detect occlusion. Christian et al. [63] proposed a novel approach to estimate 3D scene modeling using monocular scene. This approach enables multi-frame inference in tracking by detection framework using a probabilistic 3D scene model. The algorithm performs monocular 3D scene geometry estimation in real-time traffic scenes. Such system handles clutter background and occlusion. Ess et al. [64] proposed a 2D Walsh-Hadamard filter bank and depth information obtained from stereo disparity to infer traffic situations. Brostow et al. [65] suggested a novel approach to understand the traffic scene. This algorithm uses 3D point clouds to get a better 2D scene segmentation.

Tu et al. [66] explained a new model for image understanding. The proposed approach uses Markov chain Monte Carlo (MCMC) sampling method to discriminate classifiers with bottom-up generative models for 2D image understanding. Hoiem et al. [67] proposed image segmentation and object detection to deduce the object positions in 3D. In the proposed approach multi-cue combination of scene labels and object detectors allows to reinforce weak detections. Taiki et al. [68] proposed a novel approach to detect and track multiple moving objects in monocular scenes. This approach recovers 3D models separately using multiple calibrated cameras. The algorithm uses target which is a combination of moving objects and its artifacts for tracking purposes. A mixture model with semiparametric PDF generates 3D coordinates and estimates their position. Such an algorithm handles occlusion and tracks multiple objects with similar appearance. Byeon et al. [69] proposed a unique approach for the 3D model of objects. This approach uses data association method instead of appearance to generate 3D estimation. Iwashita et al. [70] suggested a level set geometry-based method to estimate the 3D coordinates. The algorithm can handle occlusion and objects that are of the same appearance. Luo et al. [71] proposed a unique geometry-based human body fitting 3D model technique. This approach can handle objects which are occluded partially with the other similar objects or background.

Xiaozhi et al. [72] explained the application of 3D scene and recognition in 3D coordinates for autonomous driving. High-quality 3D detection from monocular scene is obtained by using standard convolutional neural network pipeline. This approach has utilized the energy minimization method to place the object in 3D environment. The algorithm performance evaluation shows that it can handle challenges and generate appropriate 3D model in monocular scenes. Chen et al. [73] proposed a unique approach to exploit stereo imagery to create appropriate 3D proposals. The algorithm uses RGB-D inputs to score 3D bounding boxes using a conditional random field. Karpathy et al. [74] proposed novel RGB-D approach to estimate 3D scene. This approach uses depth information from the sensor to predict the 3D scenes through shape analysis. Koltun et al. [75] proposed an algorithm that ensembles binary segmentation models for object candidates. Joint learning of the ensembles of local and global binary conditional random fields used individual predictors to specialize in different conduct. T. Lee et al. [76] used parametric energies to propose promising object candidate. These parametric energies suggest multiple divergence regions.

Trivedi et al. [77] explained how the detection approach occurs in autonomous driving. It detects a candidate set of objects and fits a deformable part model within a box. The proposed approach uses an ensemble of model derived from visual and geometrical clusters of object instances. Pepik et al. [78] proposed a unique approach to detect a candidate set of objects. A deformable part model fits within a box. This approach extends those deformable part models to 3D linking parts across different viewpoints. Xiang et al. [79] suggested a unique approach for 3D object recognition. The algorithm uses a 3D voxel pattern (3DVP) supervised learning approach. It learns occlusion patterns to improve performance evaluation. Fidler et al. [80] proposed a novel approach to extend the deformable part-based model. The proposed model represents deformable 3D cuboid of faces and parts. The algorithm has been tested for both indoor and outdoor environments intended for 2D and 3D object detection. Vacchetti et al. [81] recommended a novel approach for online and offline 3D tracking. This approach uses jitter-free tracker that combines natural feature matching and key frames to handle camera displacement. It also makes it robust.

Peter et al. [82] proposed a novel unique geometry-based approach for object detection in 3D monocular scene. This algorithm generates geometric primitives producing predetermined location pattern in occupancy map. We can estimate object location by deconvolving the occupancy map and template similarity techniques. The same approach can be used for single and multiview monocular scenes. The proposed monocular algorithm shows significant improvement in offline compare to other state-of-the-art approaches. Kelly et al. [83] suggested a novel approach to construct 3D model using voxel feature. People and other moving objects can be modeled as a set of these voxels to decide the camera hand-off trouble. Sato et al. [84] proposed CAD-based 3D environment model which is used to extract 3D location of unknown moving foreground objects. Jain et al. [85] recommended a unique approach to obtain 3D location of every moving object using a highly calibrated camera to sense the environment and obtain the 3D locations of every unknown moving object for multiple perspective interactive video.

2.7   Object Tracking

For surveillance in both indoor and outdoor environments, tracking moving objects in video sequences is important and challenging. The development of new computer systems and high-quality cameras and the necessity for vigilant monitoring spur development and implementation of object detection tracking algorithms. If it follows certain steps, an algorithm like this would appear to be smart and vigilant. Objects in the foreground must be tracked over a period of successive frames and moving objects must be detected. Furthermore, the object's behavior must also be identified and analyzed consecutively. Various methods of object tracking are shown in the following literature.

Some of the existing general approaches for smart and intelligent object tracking are feature-based object tracking, region-based object tracking, contour-based object tracking, and model-based object tracking. Generally, the classification is based on object representation, tracker features, and image feature selection. Trackers and image features rely on the video environment, the appearance of the object, its silhouette, the number of objects, and the position and movement of the surveillance camera. Objects that follow certain geometry, such as a rectangle or ellipse, are better represented by point and primitive geometry-based approaches. Skeletal or part-based tracking models are preferable when tracking targets are complicated or uttered.

2.7.1   Feature-Based Tracking

Based on object elements, the feature-based tracking method extracts object features from the object and tracks them. It also matches those elements among images taken within consecutive frames by clustering those extracted elements. The extraction of features and correspondence of features are essential to feature-based approaches. As similar feature points are often available in successive frames, such an approach shows certain uncertainty about the feature correspondence.

An innovative survey based on features was presented by Hu et al. [86]. There are three categories in the proposed survey: global feature-based, local feature-based, and graph-based tracking methods. Color, area, perimeter, centroids, etc. are the inputs used by the global feature-based approach. Line segments, curve segments, or corner vertices of objects are used in the local feature-based approach. Vehicle tracking and traffic navigation would benefit greatly from such an approach. Dependence graphs can be used to measure geometric relationships and distances between features. Online tracking generally is not feasible due to the high computational cost of searching and matching graphs. According to Xue et al. [87], a discriminative feature of the background can be used for tracking. By applying these discriminative features, the proposed approach improved the mean shift algorithm. Yang et al. [88] proposed a unique object tracking framework based on mean shift and feature matching. In the proposed approach, feature points of template objects are extracted using a Harris detector.

A novel tracking approach based on SIFT is proposed by Rehman et al. [89] to track single and multiple objects in a variety of motion environments. In video sequences, objects are detected by background subtraction and tracked using SIFT (scale-invariant feature transformation) features. Using color as a feature and a particle filter, Fazill et al. [90] proposed an object tracking technique that combines SIFT and particle filters. Object localization and representation are handled by SIFT features, while the particle filter approximates the result of sequential estimation. A unique algorithm for tracking mean shifts and features has been presented by Bai [91]. Kalman filters are used to estimate state values, while pixels are used to create the feature space. The overall tracking approach presented by Miao et al. [92] integrates the local features of the dataset with an adaptive online boosting algorithm that uses local features. An algorithm for the tracking of mobile objects was described by Fan et al. [93]. Tracking moving objects is possible using hair-like features. Detecting features and filtering colors provide robustness and enable tracking to be recovered. Object tracking was carried out using particle filtering by Shen et al. [94]. This method can be used for nonlinear and non-Gaussian models. Mahendran et al. [95] proposed a novel object tracking approach that utilizes DML (distance matrix learning) and NN (nearest neighbor). The foreground is detected by Canny edge detectors. Due to its DML approach, such an algorithm can be applied in real time. In feature-based tracking, structures that overlap, are occluded, or are unrelated are not considered. Its combination of feature-based tracking and a different type of tracking enhances performance evaluation and addresses the above constraints.

2.7.2   Region-Based Tracking

In region-based tracking, objects are tracked based on variations of image regions relative to foreground regions. By using a simple technique called background subtraction, motion can be detected. To track objects, region-based trackers apply color, texture, motion, and spatial-temporal features. While it's computationally efficient compared with the above method, its accuracy of tracking efficiently depends on the robustness of the background model.

A novel approach to tracking single people inside of an enclosed environment was presented by Wren et al. [9]. To track various body parts, blob tracking was used. The MSS (multispectral satellite imagery) and MDL (minimum description length) algorithm that Kauth et al. [96] proposed are unique for achieving object tracking. Using the feature vector algorithm, each pixel is converted into a cluster of features called a blob. A Gaussian distribution model is applied to the background scenes by modeling each blob's spatial and color information. Video sequences were segmented using a unique supervised method by Xu et al. [97]. This approach includes estimating object motion and segmenting regions for moving objects. The object outline defined in the proposed procedure should be considered a video object. An object tracking method using backward region-based classification was proposed by Gu et al. [98]. As part of the approach, preprocessing and postprocessing are required, along with the extraction of regions, estimation of the regions, and classification of regions.

A novel approach to object tracking was presented by Hariharakrishna et al. [99]. Block motion vectors are used to track object boundaries and block motion vectors to track moving objects. In adaptive block estimation, successive frames are used to estimate motion. To detect occlusions, the dual principle is used. The tracking technique used by Andrade et al. [100] utilized descriptors derived from regions for tracking and segmentation. As object extraction can be carried out pixel by pixel with homogeneous regions in pictures, a series of images can be produced. AdaBoost-based global color features and pixel-wise tracking are used to extract an object by Wei et al. [101]. To initialize and recognize the frames, the k-means clustering algorithm is used. Bidirectional labeling can be used to achieve region-based tracking. Tracking multiple objects is made possible by a novel region-based method proposed by Kim et al. [102]. The differential image is used for tracking in the proposed approach. The proposed algorithm integrates particle filtering to handle complex environments while also providing robust tracking.

The algorithm by Khraief et al. [103] uses automatic initialization and robust background modeling to perform a unique detection tracking. Tracking is done with statistical information, while detection is done with the level set method. Detection can be combined with dynamic tracking in the proposed algorithm. Using a unique region-based approach, Varas et al. [104] tracked generic objects in a wide range of environments. A set of particle filters based on region and color is proposed. Multi-hypotheses can be handled by particle filters.

2.7.3 Contour-Based Tracking

Tracking contours analyze an object's boundaries to determine its shape and then tracks the boundaries over time. Tracking is highly dependent on how it is initiated. Due to this, automatic tracking would be very difficult. An active contour-based tracking algorithm was derived by Dokladal et al. [105]. The driver's face is tracked using feature-weighted gradients and contours. To segment an image, the gradient is computed, and the gradient is used to track the image. The contour model presented by Chen et al. [106] is based on neural fuzzy networks. For training and recognizing, we construct neural fuzzy inference networks. Histograms of silhouettes of people are projected in discrete forms.

The particle filter presented by Zhou et al. [107] is based on a unifying multi-hypothesis tracking algorithm. Integrating color features with contour information is the proposed method. For contour detection, the Sobel operator and silhouette similarity have been used. An object can be located using a color histogram and the multifusion fusion approach according to Lin et al. [108]. For the extraction of contours to track the moving objects, a region-based method can be used. Feature fusion is carried out using Harris corners for particle filters. According to Hu et al. [109], object tracking can be a very effective method. An adaptive silhouette contour evolution, color-based contour evolution, and a Markov-based dynamic model are adopted in the proposed method.

2.7.4 Model-Based Tracking

Matching scene information to the projected model is a model based tracking technique. Models are created based on prior knowledge about the object. It is usually either manual or computer vision based to deliberate projected models. Ramanan et al. [110] developed a 2D model that identifies and tracks people by tracking body parts. As the model learns from the frame set containing people's body parts, people's appearance can be modeled. Body parts are tracked in each successive frame. Poor tracking accuracy may result from bad clusters. Zhao et al. [111] proposed a new tracking method for multiple objects. Several objects can be tracked by such an algorithm in a complex environment using one camera or can be used for monocular scenes. Global and motion factors can be used to segregate people's complete motion. For tracking purposes, this method combines 3D models based on different regions.

2.7.5   Kalman Filtering

A Kalman filter computes expected values with an analysis tool and data processing algorithm. It is a mathematical process that uses a series of equations and data inputs to estimate with efficiency the actual value, position, and velocity of the object. The models can estimate past and present states as well as future states, even if the precise nature of the modeled system is unknown. Chen [112] proposed a unifying approach for tracking objects based on Kalman filters and Bhattacharya coefficients. Object positions are predicted and evaluated using both novel techniques. There are two ways to track. A kernel-based tracking method can be used to track objects in complex environments to improve tracking accuracy. A contour-based tracking method can be used after localizing the object. Wu et al. [113] developed a uniquely designed 3D tracking model. This method is capable of extracting object motion trajectories independent of the environment's complexity and unpredictability. A Kalman filter is responsible for combining both analysis techniques to enhance performance evaluation by increasing the tracking accuracy.

Using an adaptive Kalman algorithm, Li et al. [114] developed a unified object tracking approach. Moving objects are represented by corner features. In successive frames, one can estimate the parameters for the Kalman filter by looking at corner point variations. According to Peterfreund [115], a unique approach to tracking nonrigid moving objects is based on Kalman filters. When nonrigid objects do not have rigid boundaries, the active contour model simulates their velocity using Kalman filters. An optically based detection method can prevent measurements from being made and provide greater robustness in a cluttered environment.

Tracking moving objects has been introduced by Jang et al. [116]. The structural features of a moving object, the texture, the color, and the edge are incorporated into an active template dynamically. Feature energy can be minimized to achieve tracking using the Kalman filter. A Kalman filter-based tracking technique for 3D models was proposed by Fieguth et al. [117]. Kalman filters are responsible for predicting the position of the intriguing point in every next frame. Local search is responsible for recording the neighbors, and when predictions are made using 3D, the scale factor of the model needs to be considered.

2.8   Performance Evaluation

Certain metrics and measures can be used to evaluate the performance of object detection and tracking. The purpose of applying certain metrics is to compare the robustness of the proposed approach with the ground truth and other similar approaches, as well as to verify the strengths and weaknesses of various object detection and tracking approaches. The purpose of this is to establish the originality and credibility of the proposed approach.

The comparative study by Yong et al. [118] looked at how different backgrounds are modeled. Each background modeling method has been reviewed, including pixel-based, region-based, and hybrid methods. A very challenging dataset, CDnet 2014, was analyzed using eight different methods for comparison. A comparison is performed for motion segmentation and background subtraction. In addition to these features, it also compares recall, precision, FPR, FNR, and F-measure. In terms of comparing their proposed solutions with other comparable solutions, this paper is very useful for researchers.

An overview and comparison of background subtraction algorithms was presented by Yannick et al. [119]. Seven different methods of background subtraction have been compared. Each technique is compared to video and synthetic video datasets. In addition, all the techniques were also evaluated against differing constraints and a PR curve was generated for each. For researchers, this paper offers a comparison of their proposed subtraction algorithm.

An alternative approach to tracking 3D objects in monocular sequence was presented by Taiki [68]. In nature, similar objects can be tracked using this method. In a paper they propose, they compare other similar state-of-the-art methods for tracking accuracy and precision. In addition to false positives and false negatives, they also analyzed identity mismatches. Additionally, researchers may find this paper useful by comparing the accuracy and precision of their 3D monocular approaches to tracking multiple objects. H. Possengger et al. [120] proposed a unique method for tracking multiple objects in real time. They used volumetric mass densities to track moving objects in real time. It is useful to compare the proposed method with other similar approaches [53, 121] so that others can compare their proposed tracking methods.

2.9   Summary of Chapter

An overview of the various surveillance systems is presented in this chapter. As part of the literature review, we reviewed different methods for detecting and tracking objects indoors and outdoors. There are three types of background modeling: pixel-based, region-based, and hybrid. The surveillance system can be robust in the sense that it can handle dynamic environments because all approaches have limitations. Modeling techniques based on pixels are fast and accurate. There are four different algorithms available for tracking purposes. From a general literature survey, it has been found that all tracking techniques fall into the features, regions, contours, or model categories.

A number of applications include navigation using inertial navigation units, GPS, tracking objects (e.g., missiles, faces, heads, hands), stabilizing depth measurements, feature tracking, cluster tracking, composite radar, laser scanner, and stereo-camera data. In Kalman filtering, two major processes are performed. In the area of object tracking, most literature included Kalman filtering as a method of prediction and measurement, giving optimal estimates of (future) outputs. The proposed algorithm uses Kalman filtering for object tracking. For 2D foregrounds, Gaussian mixture model (GMM) is used, while for 3D foregrounds, voxels are used instead of pixels for detection. In addition to the Gaussian mixture model (GMM), we detect voxels using semiparametric PDF (SP-PDF) and MAP-EM. To improve the system's robustness, researchers have attempted to combine background modeling approaches. However, based on the literature study, it appears that the perfect resolution has not yet been established for improving performance or robustness.

According to this literature review and the problem definition, the Gaussian mixture model can be employed for two-dimensional object detection. Combining mixture models and SP-PDF can effectively detect objects in three dimensions. Kalman filtering can help attain tracking accuracy for various environments and background-foreground restrictions at the same time. Following the findings from the preceding literature, a modified Gaussian mixture model (GMM) and Kalman filtering are used to detect objects and to track objects.

References

1. Collins, R. T., Lipton, A. J., Kanade, T., Fujiyoshi, H., Duggins, D., Tsin, Y., & Wixson, L. (2000). A system for video surveillance and monitoring. In *VSAM final report, 2000* (pp. 1–68). Robotics Institute, CMU.
2. Shah, M., Javed, O., & Shafique, K. (2007). Automated visual surveillance in realistic scenarios. *IEEE Multimedia, 14*(1), 30–39.
3. Tian, Y. L., Brown, L., Hampapur, A., Lu, M., Senior, A., & Shu, C. F. (2008). IBM smart surveillance system (S3): Event-based video surveillance system with an open and extensible framework. *Machine Vision and Applications, 19*(5), 315–327.
4. Stauffer, C., & Grimson, W. E. L. (2000). Learning patterns of activity using real-time tracking. *IEEE Transactions on Pattern Analysis and Machine Intelligence, 22*(8), 747–757.
5. KaewTraKulPong, P., & Bowden, R. (2002). An improved adaptive background mixture model for real-time tracking with shadow detection. In *Video-based surveillance systems* (pp. 135–144). Springer.
6. Wang, H., & Suter, D. (2005, March). A re-evaluation of mixture of Gaussian background modeling [video signal processing applications]. In *Proceedings. (ICASSP '05). IEEE international conference on acoustics, speech, and signal processing, 2005* (Vol. 2, pp. ii–1017). IEEE.
7. Harville, M., Gordon, G., & Woodfill, J. (2001, July). Foreground segmentation using adaptive mixture models in color and depth. In *Proceedings IEEE workshop on detection and recognition of events in video* (pp. 3–11). IEEE.
8. Harville, M. (2002, May). A framework for high-level feedback to adaptive, per-pixel, mixture-of-Gaussian background models. In *European conference on computer vision* (pp. 543–560). Springer.
9. Wren, C. R., Azarbayejani, A., Darrell, T., & Pentland, A. P. (1997). Pfinder: Real-time tracking of the human body. *IEEE Transactions on Pattern Analysis and Machine Intelligence, 19*(7), 780–785.
10. Cheng, J., Yang, J., Zhou, Y., & Cui, Y. (2006). Flexible background mixture models for foreground segmentation. *Image and Vision Computing, 24*(5), 473–482.
11. Shimada, A., Arita, D., & Taniguchi, R. I. (2006, November). Dynamic control of adaptive mixture-of-Gaussian's background model. In *2006 IEEE international conference on video and signal based surveillance* (pp. 5–5). IEEE.

12. Tan, R., Huo, H., Qian, J., & Fang, T. (2006, August). Traffic video segmentation using adaptive K Gaussian mixture model. In *International workshop on intelligent computing in pattern analysis and synthesis* (pp. 125–134). Springer.

13. Carminati, L., & Benois-Pineau, J. (2005, September). Gaussian mixture classification for moving object detection in video surveillance environment. In *IEEE international conference on image processing 2005* (Vol. 3, pp. III–113). IEEE.

14. Morellas, V., Pavlidis, I., & Tsiamyrtzis, P. (2003). DETER: Detection of events for threat evaluation and recognition. *Machine Vision and Applications, 15*(1), 29–45.

15. Lee, D. S. (2004, May). Online adaptive Gaussian mixture learning for video applications. In *International workshop on statistical methods in video processing* (pp. 105–116). Springer.

16. Zhang, Y., Liang, Z., Hou, Z., Wang, H., & Tan, M. (2005, December). An adaptive mixture Gaussian background model with online background reconstruction and adjustable foreground mergence time for motion segmentation. In *2005 IEEE international conference on industrial technology* (pp. 23–27). IEEE.

17. Amintoosi, M., Farbiz, F., Fathy, M., Analoui, M., & Mozayani, N. (2007, April). QR decomposition-based algorithm for background subtraction. In *2007 IEEE international conference on acoustics, speech and signal processing-ICASSP'07* (Vol. 1, pp. I–1093). IEEE.

18. White, B., & Shah, M. (2007, July). Automatically tuning background subtraction parameters using particle swarm optimization. In *2007 IEEE international conference on multimedia and expo* (pp. 1826–1829). IEEE.

19. Javed, O., Shafique, K., & Shah, M. (2002, December). A hierarchical approach to robust background subtraction using color and gradient information. In *Workshop on motion and video computing, 2002. Proceedings* (pp. 22–27). IEEE.

20. Zang, Q., & Klette, R. (2004, August). Robust background subtraction and maintenance. In *Proceedings of the 17th international conference on pattern recognition, 2004. ICPR 2004* (Vol. 2, pp. 90–93). IEEE.

21. Butler, D., Sridharan, S., & Bove, V. J. (2003, April). Real-time adaptive background segmentation. In *2003 IEEE international conference on acoustics, speech, and signal processing, 2003. Proceedings. (ICASSP'03)* (Vol. 3, pp. III–349). IEEE.

22. Zivkovic, Z., & van der Heijden, F. (2004). Recursive unsupervised learning of finite mixture models. *IEEE Transactions on Pattern Analysis and Machine Intelligence, 26*(5), 651–656.

23. Lee, D. S. (2005). Effective Gaussian mixture learning for video background subtraction. *IEEE Transactions on Pattern Analysis and Machine Intelligence, 27*(5), 827–832.

24. Landabaso, J. L., Pardàs, M., & Xu, L. Q. (2005, March). Hierarchical representation of scenes using activity information. In *Proceedings. (ICASSP'05). IEEE international conference on acoustics, speech, and signal processing, 2005* (Vol. 2, pp. ii–677). IEEE.

25. Schindler, K., & Wang, H. (2006, January). Smooth foreground-background segmentation for video processing. In *Asian conference on computer vision* (pp. 581–590). Springer.

26. Ma, R., Li, L., Huang, W., & Tian, Q. (2004, December). On pixel count based crowd density estimation for visual surveillance. In *IEEE conference on cybernetics and intelligent systems, 2004* (Vol. 1, pp. 170–173). IEEE.

27. Chan, A. B., Liang, Z. S. J., & Vasconcelos, N. (2008, June). Privacy preserving crowd monitoring: Counting people without people models or tracking. In *2008 IEEE conference on computer vision and pattern recognition* (pp. 1–7). IEEE.

28. Lipton, A. J., Fujiyoshi, H., & Patil, R. S. (1998, October). Moving target classification and tracking from real-time video. In *Proceedings fourth IEEE workshop on applications of computer vision. WACV'98 (Cat. No. 98EX201)* (pp. 8–14). IEEE.

29. Jacques, J. C. S., Jung, C. R., & Musse, S. R. (2005, October). Background subtraction and shadow detection in grayscale video sequences. In *XVIII Brazilian symposium on computer graphics and image processing (SIBGRAPI'05)* (pp. 189–196). IEEE.

30. Haritaoglu, I., Harwood, D., & Davis, L. S. (2000). W/sup 4: Real-time surveillance of people and their activities. *IEEE Transactions on Pattern Analysis and Machine Intelligence, 22*(8), 809–830.

31. Kim, K., Chalidabhongse, T. H., Harwood, D., & Davis, L. (2004, October). Background modeling and subtraction by codebook construction. In *2004 international conference on image processing, 2004. ICIP '04* (Vol. 5, pp. 3061–3064). IEEE.
32. Kim, K., Chalidabhongse, T. H., Harwood, D., & Davis, L. (2005). Real-time foreground–background segmentation using codebook model. *Real-Time Imaging, 11*(3), 172–185.
33. Oliver, N. M., Rosario, B., & Pentland, A. P. (2000). A Bayesian computer vision system for modeling human interactions. *IEEE Transactions on Pattern Analysis and Machine Intelligence, 22*(8), 831–843.
34. Xu, Z., Shi, P., & Gu, I. Y. H. (2006, November). An eigenbackground subtraction method using recursive error compensation. In *Pacific-Rim conference on multimedia* (pp. 779–787). Springer.
35. Elgammal, A., Harwood, D., & Davis, L. (2000, June). Non-parametric model for background subtraction. In *European conference on computer vision* (pp. 751–767). Springer.
36. Russell, D., & Gong, S. (2005, September). A highly efficient block-based dynamic background model. In *IEEE conference on advanced video and signal based surveillance, 2005* (pp. 417–422). IEEE.
37. Heikkila, M., & Pietikainen, M. (2006). A texture-based method for modeling the background and detecting moving objects. *IEEE Transactions on Pattern Analysis and Machine Intelligence, 28*(4), 657–662.
38. Seki, M., Wada, T., Fujiwara, H., & Sumi, K. (2003, June). Background subtraction based on cooccurrence of image variations. In *2003 IEEE Computer Society Conference on Computer Vision and Pattern Recognition, 2003. Proceedings* (Vol. 2, p. II–II). IEEE.
39. Liu, W. C., Lin, S. Z., Yang, M. H., & Huang, C. R. (2013, November). Real-time binary descriptor-based background modeling. In *2013 2nd IAPR Asian conference on pattern recognition* (pp. 722–726). IEEE.
40. Travakkoli, A., Nicolescu, M., & Bebis, G. (2006, March). Automatic statistical object detection for visual surveillance. In *2006 IEEE southwest symposium on image analysis and interpretation* (pp. 144–148). IEEE.
41. Tanaka, T., Shimada, A., Arita, D., & Taniguchi, R. I. (2007, November). Non-parametric background and shadow modeling for object detection. In *Asian conference on computer vision* (pp. 159–168). Springer.
42. Mao, Y. F., & Shi, P. F. (2005). Diversity sampling-based kernel density estimation for background modeling. *Journal of Shanghai University (English Edition), 9*(6), 506–509.
43. Ianasi, C., Gui, V., Toma, C. I., & Pescaru, D. (2005). A fast algorithm for background tracking in video surveillance, using nonparametric kernel density estimation. *Facta Universitatis-Series: Electronics and Energetics, 18*(1), 127–144.
44. Patwardhan, K., Sapiro, G., & Morellas, V. (2008). Robust foreground detection in video using pixel layers. *IEEE Transactions on Pattern Analysis and Machine Intelligence, 30*(4), 746–751.
45. Abdelkader, M. F., Chellappa, R., Zheng, Q., & Chan, A. L. (2006, January). Integrated motion detection and tracking for visual surveillance. In *Fourth IEEE international conference on computer vision systems (ICVS '06)* (pp. 28–28). IEEE.
46. Grabner, H., & Bischof, H. (2006, June). On-line boosting and vision. In *2006 IEEE computer society conference on computer vision and pattern recognition (CVPR '06)* (Vol. 1, pp. 260–267). IEEE.
47. Toyama, K., Krumm, J., Brumitt, B., & Meyers, B. (1999, September). Wallflower: Principles and practice of background maintenance. In *Proceedings of the seventh IEEE international conference on computer vision* (Vol. 1, pp. 255–261). IEEE.
48. Koller, D., Weber, J., & Malik, J. (1994, May). Robust multiple car tracking with occlusion reasoning. In *European conference on computer vision* (pp. 189–196). Springer.
49. Marana, A. N., Velastin, S. A., Costa, L. D. F., & Lotufo, R. A. (1998). Automatic estimation of crowd density using texture. *Safety Science, 28*(3), 165–175.

50. Wang, H., & Suter, D. (2007). A consensus-based method for tracking: Modelling background scenario and foreground appearance. *Pattern Recognition, 40*(3), 1091–1105.
51. Maddalena, L., & Petrosino, A. (2008). A self-organizing approach to background subtraction for visual surveillance applications. *IEEE Transactions on Image Processing, 17*(7), 1168–1177.
52. Barnich, O., & Van Droogenbroeck, M. (2009, April). ViBe: A powerful random technique to estimate the background in video sequences. In *2009 IEEE international conference on acoustics, speech, and signal processing* (pp. 945–948). IEEE.
53. Hofmann, M., Tiefenbacher, P., & Rigoll, G. (2012, June). Background segmentation with feedback: The pixel-based adaptive segmenter. In *2012 IEEE computer society conference on computer vision and pattern recognition workshops* (pp. 38–43). IEEE.
54. DeMenthon, D. F., & Davis, L. S. (1995). Model-based object pose in 25 lines of code. *International Journal of Computer Vision, 15*(1–2), 123–141.
55. Huang, S. S., Fu, L. C., & Hsiao, P. Y. (2009). Region-level motion-based foreground segmentation under a Bayesian network. *IEEE Transactions on Circuits and Systems for Video Technology, 19*(4), 522–532.
56. Tsai, T. H., Lin, C. Y., & Li, S. Y. (2012). Algorithm and architecture design of human–machine interaction in foreground object detection with dynamic scene. *IEEE Transactions on Circuits and Systems for Video Technology, 23*(1), 15–29.
57. Najafi, H., Genc, Y., & Navab, N. (2006, January). Fusion of 3D and appearance models for fast object detection and pose estimation. In *Asian conference on computer vision* (pp. 415–426). Springer.
58. Crivellaro, A., Rad, M., Verdie, Y., Moo Yi, K., Fua, P., & Lepetit, V. (2015). A novel representation of parts for accurate 3D object detection and tracking in monocular images. In *Proceedings of the IEEE international conference on computer vision* (pp. 4391–4399). IEEE.
59. Xiang, Y., Song, C., Mottaghi, R., & Savarese, S. (2014, September). Monocular multiview object tracking with 3D aspect parts. In *European conference on computer vision* (pp. 220–235). Springer.
60. Brachmann, E., Krull, A., Michel, F., Gumhold, S., Shotton, J., & Rother, C. (2014, September). Learning 6D object pose estimation using 3D object coordinates. In *European conference on computer vision* (pp. 536–551). Springer.
61. Chliveros, G., Pateraki, M., & Trahanias, P. (2013, July). Robust multi-hypothesis 3D object poses tracking. In *International conference on computer vision systems* (pp. 234–243). Springer.
62. Tejani, A., Tang, D., Kouskouridas, R., & Kim, T. K. (2014, September). Latent class Hough forests for 3D object detection and poses estimation. In *European conference on computer vision* (pp. 462–477). Springer.
63. Wojek, C., Roth, S., Schindler, K., & Schiele, B. (2010, September). Monocular 3D scene modeling and inference: Understanding multi-object traffic scenes. In *European conference on computer vision* (pp. 467–481). Springer.
64. Ess, A., Müller, T., Grabner, H., & Van Gool, L. (2009, September). Segmentation-based urban traffic scene understanding. In *BMVC* (Vol. 1, p. 2). The British Machine Vision Association and Society for Pattern Recognition.
65. Brostow, G. J., Shotton, J., Fauqueur, J., & Cipolla, R. (2008, October). Segmentation and recognition using structure from motion point clouds. In *European conference on computer vision* (pp. 44–57). Springer.
66. Tu, Z., Chen, X., Yuille, A. L., & Zhu, S. C. (2005). Image parsing: Unifying segmentation, detection, and recognition. *International Journal of Computer Vision, 63*(2), 113–140.
67. Hoiem, D., Efros, A. A., & Hebert, M. (2008). Putting objects in perspective. *International Journal of Computer Vision, 80*(1), 3–15.
68. Sekii, T. (2016). Robust, real-time 3D tracking of multiple objects with similar appearances. In *Proceedings of the IEEE conference on computer vision and pattern recognition* (pp. 4275–4283). IEEE.

69. Byeon, M., Oh, S., Kim, K., Yoo, H. J., & Choi, J. Y. (2015). Efficient spatio-temporal data association using multidimensional assignment in multi-camera multi-target tracking. In *BMVC* (pp. 1–68). The British Machine Vision Association and Society for Pattern Recognition.
70. Iwashita, Y., Kurazume, R., Hasegawa, T., & Hara, K. (2006, May). Robust motion capture system against target occlusion using fast level set method. In *Proceedings 2006 IEEE international conference on robotics and automation, 2006. ICRA 2006* (pp. 168–174). IEEE.
71. Luo, X., Berendsen, B., Tan, R. T., & Veltkamp, R. C. (2010, August). Human pose estimation for multiple persons based on volume reconstruction. In *2010 20th international conference on pattern recognition* (pp. 3591–3594). IEEE.
72. Chen, X., Kundu, K., Zhang, Z., Ma, H., Fidler, S., & Urtasun, R. (2016). Monocular 3D object detection for autonomous driving. In *Proceedings of the IEEE conference on computer vision and pattern recognition* (pp. 2147–2156). IEEE.
73. Chen, X., Kundu, K., Zhu, Y., Berneshawi, A. G., Ma, H., Fidler, S., & Urtasun, R. (2015). 3D object proposals for accurate object class detection. In *Advances in neural information processing systems* (pp. 424–432). IEEE.
74. Karpathy, A., Miller, S., & Fei-Fei, L. (2013, May). Object discovery in 3D scenes via shape analysis. In *2013 IEEE international conference on robotics and automation* (pp. 2088–2095). IEEE.
75. Krahenbuhl, P., & Koltun, V. (2015). Learning to propose objects. In *Proceedings of the IEEE conference on computer vision and pattern recognition* (pp. 1574–1582). IEEE.
76. Lee, T., Fidler, S., & Dickinson, S. (2015). A learning framework for generating region proposals with mid-level cues. In *International conference on computer vision* (Vol. 2). IEEE.
77. Ohn-Bar, E., & Trivedi, M. M. (2015). Learning to detect vehicles by clustering appearance patterns. *IEEE Transactions on Intelligent Transportation Systems, 16*(5), 2511–2521.
78. Pepik, B., Stark, M., Gehler, P., & Schiele, B. (2015). Multi-view and 3D deformable part models. *IEEE Transactions on Pattern Analysis and Machine Intelligence, 37*(11), 2232–2245.
79. Xiang, Y., Choi, W., Lin, Y., & Savarese, S. (2015). Data-driven 3D voxel patterns for object category recognition. In *Proceedings of the IEEE conference on computer vision and pattern recognition* (pp. 1903–1911). IEEE.
80. Leibe, B., Cornelis, N., Cornelis, K., & Van Gool, L. (2007, June). Dynamic 3D scene analysis from a moving vehicle. In *2007 IEEE conference on computer vision and pattern recognition* (pp. 1–8). IEEE.
81. Vacchetti, L., Lepetit, V., & Fua, P. (2004). Stable real-time 3D tracking using online and offline information. *IEEE Transactions on Pattern Analysis and Machine Intelligence, 26*(10), 1385–1391.
82. Carr, P., Sheikh, Y., & Matthews, I. (2012, October). Monocular object detection using 3D geometric primitives. In *European conference on computer vision* (pp. 864–878). Springer.
83. Kelly, P. H., Katkere, A., Kuramura, D. Y., Moezzi, S., & Chatterjee, S. (1995, January). An architecture for multiple perspective interactive video. In *Proceedings of the third ACM international conference on multimedia* (pp. 201–212). ACM.
84. Sato, K., Maeda, T., Kato, H., & Inokuchi, S. (1994, February). CAD-based object tracking with distributed monocular camera for security monitoring. In *Proceedings of 1994 IEEE 2nd CAD-based vision workshop* (pp. 291–297). IEEE.
85. Jain, R., & Wakimoto, K. (1995, May). Multiple perspective interactive video. In *Proceedings of the international conference on multimedia computing and systems* (pp. 202–211). IEEE.
86. Hu, W., Tan, T., Wang, L., & Maybank, S. (2004). A survey on visual surveillance of object motion and behaviors. *IEEE Transactions on Systems, Man, and Cybernetics, Part C (Applications and Reviews), 34*(3), 334–352.
87. Xue, C., Zhu, M., & Chen, A. H. (2008, December). A discriminative feature-based mean-shift algorithm for object tracking. In *2008 IEEE international symposium on knowledge acquisition and modeling workshop* (pp. 217–220). IEEE.

88. Yang, W., Li, J., Liu, J., & Shi, D. (2009, August). A novel layered object tracking algorithm for forward-looking infrared imagery based on mean shift and feature matching. In *2009 2nd IEEE international conference on computer science and information technology* (pp. 188–191). IEEE.

89. Rahman, M. S., Saha, A., & Khanum, S. (2009, November). Multi-object tracking in video sequences based on background subtraction and sift feature matching. In *2009 Fourth international conference on computer sciences and convergence information technology* (pp. 457–462). IEEE.

90. Fazli, S., Pour, H. M., & Bouzari, H. (2009, December). Particle filter-based object tracking with sift and color feature. In *2009 Second international conference on machine vision* (pp. 89–93). IEEE.

91. Bai, K. J. (2010, June). A new object tracking algorithm based on mean shift in 4-D state space and on-line feature selection. In *2010 Third international conference on information and computing* (Vol. 1, pp. 39–42). IEEE.

92. Miao, Q., Wang, G., Lin, X., Wang, Y., Shi, C., & Liao, C. (2010, September). Scale and rotation invariant feature-based object tracking via modified on-line boosting. In *2010 IEEE international conference on image processing* (pp. 3929–3932). IEEE.

93. Fan, L., Riihimaki, M., & Kunttu, I. (2010, September). A feature-based object tracking approach for real time image processing on mobile devices. In *2010 IEEE international conference on image processing* (pp. 3921–3924). IEEE.

94. Shen, H. Y., Sun, S. F., Ma, X. B., Xu, Y. C., & Lei, B. J. (2012, July). Comparative study of color feature for particle filter based object tracking. In *2012 International conference on machine learning and cybernetics* (Vol. 3, pp. 1104–1110). IEEE.

95. Mahendran, S., Vaithiyanathan, D., & Seshasayanan, R. (2013, April). Object tracking system based on invariant features. In *2013 International conference on communication and signal processing* (pp. 1138–1142). IEEE.

96. Kauth, R. J., Pentland, A. P., & Thomas, G. S. (1977, January). Blob: An unsupervised clustering approach to spatial preprocessing of MSS imagery. In *Proceedings of the 11th international symposium on remote sensing of environment* (Vol. 2). IEEE.

97. Xu, D., Hwang, J. N., & Yu, J. (1999, September). An accurate region based object tracking for video sequences. In *1999 IEEE third workshop on multimedia signal processing (Cat. No. 99TH8451)* (pp. 271–276). IEEE.

98. Gu, C., & Lee, M. C. (1998, October). Semantic video object tracking using region-based classification. In *Proceedings 1998 international conference on image processing. ICIP98 (Cat. No. 98CB36269)* (pp. 643–647). IEEE.

99. Hariharakrishnan, K., & Schonfeld, D. (2005). Fast object tracking using adaptive block matching. *IEEE Transactions on Multimedia, 7*(5), 853–859.

100. Andrade, E. L., Woods, J. C., Khan, E., & Ghanbari, M. (2005). Region-based analysis and retrieval for tracking of semantic objects and provision of augmented information in interactive sport scenes. *IEEE Transactions on Multimedia, 7*(6), 1084–1096.

101. Wei, F. T., Chou, S. T., & Lin, C. W. (2008, May). A region-based object tracking scheme using AdaBoost-based feature selection. In *2008 IEEE international symposium on circuits and systems* (pp. 2753–2756). IEEE.

102. Kim, H. B., & Sim, K. B. (2010, October). A particular object tracking in an environment of multiple moving objects. In *ICCAS 2010* (pp. 1053–1056). IEEE.

103. Khraief, C., Bourouis, S., & Hamrouni, K. (2012, May). Unsupervised video objects detection and tracking using region-based level-set. In *2012 International conference on multimedia computing and systems* (pp. 201–206). IEEE.

104. Varas, D., & Marques, F. (2012). A region-based particle filter for generic object tracking and segmentation. In *2012 19th IEEE international conference on image processing* (pp. 1333–1336). IEEE.

105. Dokladal, P., Enficiaud, R., & Dejnozkova, E. (2004, May). Contour-based object tracking with gradient-based contour attraction field. In *2004 IEEE international conference on acoustics, speech, and signal processing* (Vol. 3, pp. iii–17). IEEE.
106. Chen, T. (2009, July). Object tracking based on active contour model by neural fuzzy network. In *2009 IITA international conference on control, automation and systems engineering (case 2009)* (pp. 570–574). IEEE.
107. Pu, B., Zhou, F., & Bai, X. (2011, October). Particle filter based on color feature with contour information adaptively integrated for object tracking. In *2011 Fourth international symposium on computational intelligence and design* (Vol. 2, pp. 359–362). IEEE.
108. Lu, X., Song, L., Yu, S., & Ling, N. (2012, July). Object contour tracking using multi-feature fusion-based particle filter. In *2012 7th IEEE conference on industrial electronics and applications (ICIEA)* (pp. 237–242). IEEE.
109. Hu, W., Zhou, X., Li, W., Luo, W., Zhang, X., & Maybank, S. (2012). Active contour-based visual tracking by integrating colors, shapes, and motions. *IEEE Transactions on Image Processing, 22*(5), 1778–1792.
110. Ramanan, D., Forsyth, D. A., & Zisserman, A. (2005, June). Strike a pose: Tracking people by finding stylized poses. In *2005 IEEE computer society conference on computer vision and pattern recognition (CVPR'05)* (Vol. 1, pp. 271–278). IEEE.
111. Zhao, T., & Nevatia, R. (2004). Tracking multiple humans in complex situations. *IEEE Transactions on Pattern Analysis and Machine Intelligence, 26*(9), 1208–1221.
112. Chen, Q., Sun, Q. S., Heng, P. A., & Xia, D. S. (2010). Two-stage object tracking method based on kernel and active contour. *IEEE Transactions on Circuits and Systems for Video Technology, 20*(4), 605–609.
113. Wu, X., Mao, X., Chen, L., & Compare, A. (2013, August). Combined motion and region-based 3D tracking in active depth image sequence. In *2013 IEEE international conference on green computing and communications and IEEE internet of things and IEEE cyber, physical and social computing* (pp. 1734–1739). IEEE.
114. Li, N., Liu, L., & Xu, D. (2008, October). Corner feature-based object tracking using adaptive Kalman filter. In *2008 9th international conference on signal processing* (pp. 1432–1435). IEEE.
115. Peterfreund, N. (1999). Robust tracking of position and velocity with Kalman snakes. *IEEE Transactions on Pattern Analysis and Machine Intelligence, 21*(6), 564–569.
116. Jang, D. S., & Choi, H. I. (2000). Active models for tracking moving objects. *Pattern Recognition, 33*(7), 1135–1146.
117. Fieguth, P., & Terzopoulos, D. (1997, June). Color-based tracking of heads and other mobile objects at video frame rates. In *Proceedings of IEEE computer society conference on computer vision and pattern recognition* (pp. 21–27). IEEE.
118. Xu, Y., Dong, J., Zhang, B., & Xu, D. (2016). Background modeling methods in video analysis: A review and comparative evaluation. *CAAI Transactions on Intelligence Technology, 1*(1), 43–60.
119. Benezeth, Y., Jodoin, P. M., Emile, B., Laurent, H., & Rosenberger, C. (2010). Comparative study of background subtraction algorithms. *Journal of Electronic Imaging, 19*(3), 033003.
120. Possegger, H., Sternig, S., Mauthner, T., Roth, P. M., & Bischof, H. (2013). Robust real-time tracking of multiple objects by volumetric mass densities. In *Proceedings of the IEEE conference on computer vision and pattern recognition* (pp. 2395–2402). IEEE.
121. Berclaz, J., Fleuret, F., Turetken, E., & Fua, P. (2011). Multiple object tracking using k-shortest paths optimization. *IEEE Transactions on Pattern Analysis and Machine Intelligence, 33*(9), 1806–1819.

Chapter 3
Background Modeling

3.1 Introduction

Background modeling is a method of representing the background using models. A model's ability to adapt to a univariate or multivariate background is determined by this factor. Foreground or moving object detection is generally done with the use of background modeling for a variety of applications including indoor and outdoor video surveillance, optical motion, multimedia, robot navigation, traffic monitoring, and automatic driver assistance. Primitive video surveillance involves segmenting out the moving or foreground objects from every frame of the input video. For the tracking system to be reliable, it must be capable of estimating and subtracting background radiation effectively and robustly. It is not always necessary to establish a background model and to detect foreground objects in computer vision applications. The foreground is detected through change detection and the background is modeled by temporal variations. Video on mobile devices and the Internet has no static background, and such videos suffer from challenges. Researchers face a challenging task when it comes to detecting moving objects. In addition to handling dynamic and clutter backgrounds, indoor and outdoor video sequences, variations in light, etc., the model's efficiency is dependent on how well it handles different challenges.

3.2 Background Modeling

For modeling background, it is easiest to obtain a background image that doesn't include any foreground. Background modeling is primarily concerned with segmenting moving objects from non-static and static backgrounds. It is a challenging task to model video sequences' background when there are no available

N. Ghedia et al., *Moving Objects Detection Using Machine Learning*, SpringerBriefs in Electrical and Computer Engineering, https://doi.org/10.1007/978-3-030-90910-9_3

Table 3.1 Traditional background modeling approaches

Background approach	Background category	Background subcategory
Traditional approaches	Basic	Histogram, median, and average
	Statistical	Support vector, subspace learning
	Cluster	k-means, codebook, sequential clustering
	Neural	Regression and multi-valued neural, competitive neural
	Estimation	Wiener filtering, Kalman filtering, Chebyshev filtering

Table 3.2 Recent background modeling approaches

Background approach	Background category	Background subcategory
Recent approaches	Statistical	Gaussian mixture model, hybrid and nonparametric model, multi-kernel model
	Fuzzy	Fuzzy—background, foreground detection, background maintenance, feature, and postprocessing
	Discriminative and mixed subspace learning	Discriminative subspace, mixed subspace
	Robust subspace	Principal component analysis (PCA) and robust principal component analysis (RPCA), RPCA—sparse, outliers, sparsity
	Subspace tracking	Grassmannian robust adaptive subspace tracking algorithm (GRASTA)
	Low-rank minimization	Contiguous outliers, direct robust matrix factorization, probability robust matrix factorization
	Sparse	Compressive sensing, structured sparsity, dynamic group sparsity
	Domain transform	Fast Fourier transform (FFT), discrete cosine transform (DCT), wavelet transform (WT)

backgrounds, objects are being introduced and removed, and illumination varies both indoors and outdoors. Parametric and nonparametric models can be classified as part of this background modeling. In some cases, the approach may be divided into pixels, regions, and hybrids [1]. Recursive and non-recursive approaches to background modeling are also possible [2]. The term has also been used for the unimodal, bimodal, and multimodal modes of transportation. Generally, it can be divided into two categories: traditional and recent. Based on each of the above categories, Tables 3.1 and 3.2 show the complete traditional and recent approaches to modeling the background. Figure 3.1 illustrates the simple background modeling technique.

In Fig. 3.1, each background pixel is modeled by the background model to detect foreground objects. Backgrounds are usually dynamic and suffer from various limitations. From a very challenging and dynamic background, Fig. 3.1 illustrates an exact technique for developing the background. A robust background model is also developed by pre- and postprocessing, which also improves comparative

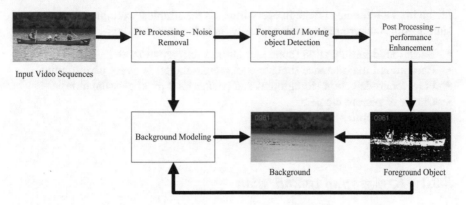

Fig. 3.1 Simple background modeling technique

performance evaluation. The background modeling technique requires low computational complexity to be used for real-time applications such as video surveillance and automated driver assistance. The foreground model should also consider the transient stops of the background objects.

At the same time, the background model is sufficiently sensitive to detect the moving objects or foreground objects despite all the environmental constraints such as changing background environments, dynamic backgrounds, and clutter backdrops. Object positions and foregrounds can be determined by two methods. These are:

- Background Subtraction: It compares every pixel with a background to determine whether it belongs to the foreground or the background. As a result, the computational complexity is low, and it is particularly useful where the background is stationary.
- Optical Flow: To detect moving or foreground objects, compare consecutive frames to estimate movement of pixels (flow vectors). Even from moving cameras, it can detect moving objects. A real-time surveillance system is not feasible due to high computational complexity.

3.3 Background Subtraction

Foreground objects are detected using a background subtraction technique. As a result of this process, the foreground moving objects can be detected by "subtracting" the background from the video sequences. If one knows a priori what the background pixels are, background subtraction is the simplest technique. A background subtraction approach can be used in a variety of applications, including video surveillance, vehicle navigation, and traffic monitoring. This system performs well in both an indoor and outdoor environment. Subtracting the current image from a reference image in every frame is the simplest example of background subtraction. Background models typically include these sections:

- Background initialization (construct initial background model)
- Background maintenance (update background model in every pixel/frame)
- Foreground detection (foreground and background pixel classification)
- Choice of picture element
- Choice of features

3.3.1   Background Initialization

To create a video surveillance system background model, a fixed number of video frames is used. A video sequence's background may not be visible in some instances. Statistical, fuzzy, PCA, etc. approaches may be used to initialize background models. Our assumption is that a model can be initialized by certain frames that do not contain moving objects. Due to the dynamic backgrounds and clutter of outdoor surveillance, it is not possible to make such an assumption in real time. Foreground segmentation using background subtraction is shown in Fig. 3.2. Foreground object motion can be detected with a simple background subtraction:

$$I_{FG} = |I_{CF} - I_{BF}| > T_{FG} \qquad (3.1)$$

where I_{FG} = final foreground object, I_{CF} = current input frame, I_{BF} = background frame, and T_{FG} = foreground threshold. I_{BF} may be a fixed number or dynamic. Background frames depend on the complexity of the background.

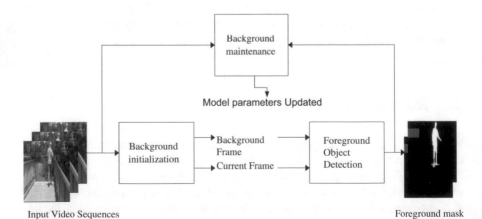

Fig. 3.2 Simple background subtraction approach

An especially challenging task is to use a sequence of frames where many initial frames contain moving objects to obtain a background model. The following techniques can be used for model initialization:

- There is a requirement for N initial frames.
- It is necessary to know how many N frames are required.
- The process continues until all the background has been obtained after N unknown frames have been generated.

As well as background initialization, it also depends on the complexity and statistics of the background models. For background initialization, the following approaches are available.

- *Temporal Frame Differencing*

It is the simplest of all background techniques. This technique is also called temporal frame differencing. The previous video frame can be used to estimate the background model. It is possible to detect foreground objects by comparing the current frame with the input frame. Calculate the temporal frame difference I_{FD} as follows:

$$T_{FD} = |I_t - I_{t-1}| \qquad (3.2)$$

where I_t is the frame pixel intensity at time t. The foreground binary mask FG can be calculated by comparing T_{FD} to the foreground threshold I_{FG} which is determined empirically, statically, or adaptively, and if difference is more than a specific threshold, then a measured pixel is to be considered as a moving or foreground pixel.

$$FG = \begin{cases} 1, & T_{FD} > T_{FG} \\ 0, & \text{otherwise} \end{cases} \qquad (3.3)$$

If $T_{FD} < T_{FG}$ then, in this case, every pixel should be treated as if there was no motion detected, and the current pixel should be treated as a background forever. Finally, a background model is established by counting the number of consecutive frames. In such a method, objects in the foreground are wrongly classified as backgrounds if they are connected to the background, and if they remain stationary, the objects will be classified as backgrounds if they are continuously moving. Background can be calculated with a single time frame difference, so it is not affected by transient stops. Noise from the camera and variations in illumination will not be handled by it. In segmentation, uniform distribution of color is also important. Large silhouettes and uniform color distributions can't be detected using frame differencing.

- *Average Filtering*

Background models are initialized using a selective averaging method. In the average filtering approach, the video frames are averaged over time to establish the background model. This method assumes that the moving objects are removed from the sequence starting frames, and that the foreground objects are available in transients. Initialize the model as follows:

$$\mathrm{BM_I} = \frac{\sum_{k=1}^{N} I_k}{N} \tag{3.4}$$

where $\mathrm{BM_I}$ is the background model pixel intensity, I_k is the pixel intensity of the kth current frame, and N is the number of frames required to develop background model. The number of frames N depends on background environment. The average filtering process works with mono-modal backgrounds, but cannot detect objects in crowded scenes, backgrounds with variations in illumination, or water twinkling on a bimodal background.

- *Median Filtering*

The median value of all pixels can be used to initialize or estimate the background. Each pixel position is stored in a memory. From the median for complete memory, the motion can be predicted by half. In a frame, a stationary object will use half of its total memory length to become the background.

3.3.2   *Background Maintenance*

To adapt to the changes that occurred in the video sequences, background maintenance is required. Every frame and pixel must be achieved and it's a learning process. The maintenance component can update the background based on the previous background, the foreground binary mask, and the current frame. Updating the background model is also necessary if a moving object remains stationary for a prolonged period. If the same incorporated moving object is again visible, it must be updated as a foreground object. A background can be maintained in every video frame during such an update. Generally, three types of maintenance are available:

- *Blind Background Maintenance*

According to this maintenance scheme, all background pixels are updated uniformly. Moreover, all background pixels are updated uniformly.

$$\mathrm{BM}_{t+1} = \mathrm{BM}_t(1 - \alpha) + \alpha I_t \tag{3.5}$$

where α is the learning rate which depends on the complexity of the background and BM_t and I_t are the background and current image at time t, respectively.

Using the learning rate to classify foreground and background, the new background is computed using the same learning rate, but this infects the image of the background.

• *Selective Background Maintenance*

A selective background maintenance scheme is proposed in the literature as a solution to the problem of blind maintenance. To classify foreground objects and backgrounds, two different learning rates are used instead of one. Our background model updates very quickly when we use the different learning parameters, and it can also accommodate several constraints like clutter and dynamic background effectively.

$$BM_{t+1} = BM_t(1 - \alpha) + \alpha I_t \quad \text{if pixel belongs to the background} \qquad (3.6)$$

$$BM_{t+1} = BM_t(1 - \beta) + \beta I_t \quad \text{if pixel belongs to the foreground} \qquad (3.7)$$

Again, such a scheme is dependent on the learning rate and value. Foreground-background classification will be uncertain due to different learning rates. Adaptation speed is determined by the learning rate. A model's ability to adapt to lighting changes and the time needed to accommodate stationary moving objects within a background model are also determined by this factor. Learning rates are associated with a variety of challenges and exhibit a variety of characteristics.

• *Fuzzy Adaptive Background Maintenance*

It is possible to overcome the uncertainty of classification using a fuzzy adaptive background maintenance scheme. To maintain and update the background model, it graduates the update rule based on the foreground detection result. The proposed scheme can detect foreground objects under all possible conditions. During background maintenance, a model is updated when there are sudden or gradual changes in each frame. If no significant changes in parameters are apparent in a pixel or frame, then updating and maintaining the background model is not needed

3.3.3 Foreground Detection

Based on certain criteria, the foreground detection technique classifies foreground objects from background objects. From the video sequences, the foreground objects usually can be computed by dividing the current frame by the background frame. There are static, predefined, adaptive, or both thresholds for detecting foreground objects. Several approaches can be used to detect motion in a video sequence, including intensity, region, texture, motion, and edge. The background model is then investigated and maintained using the foreground binary mask.

3.3.4 Picture Element

Different choices for picture elements are pixels, blocks, and clusters. Background models are initialized and maintained by selecting a picture element. The robustness of a system is determined by its size. Pixels are less prone to noise than clusters and regions.

3.3.5 Features

It is necessary to incorporate multiple features in the background modeling process to develop a robust model. The following are some of the common features used in computer vision: spatial features (edge, texture), spectral features (color), and temporal features (motion). In addition to their different properties, these features are used to address different constraints like dynamic background and motion changes, illumination variations, and clutter. In the foreground and background, color features cannot robustly handle illumination variations. Objects under changing illumination conditions and shadows can be detected by the edge and texture features. By combining these three features, the system is robust against clutter and dynamic backgrounds. Typically, features are chosen by researchers based on the challenges and constraints they face in a particular situation. Additionally, the statistical background subtraction method provides better robustness and the ability to handle complex and critical situations compared to both traditional and recent background modeling. Additionally, the statistical background subtraction method provides better robustness and the ability to handle complex and critical situations compared to both traditional and recent background modeling.

3.4 Modified Gaussian Mixture Model

The multimodal backgrounds are modeled using a Gaussian mixture model combining Gaussian and Gaussian matrices. Clutter and dynamic backgrounds are handled by it. Mixture of Gaussian was introduced by Stauffer et al. [3]. By using a mixture of normal distributions, each pixel in the video frame is represented in a way that adjusts for the multidimensionality of the background. The Gaussian mixture model is considered suitable for detecting foreground by modeling the background and subtracting it from the current frame; such an operation can be performed pixel by pixel, rather than by a region-based approach.

- *Univariate Gaussian Distribution*

A univariate model is commonly called a single Gaussian. A Gaussian probability density function is used to model the background in this approach. During

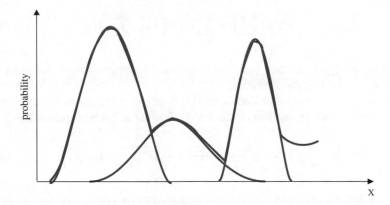

Fig. 3.3 Multivariate Gaussian distribution for $k = 3$

background model maintenance, the single Gaussian average is updated with the running average. Gaussian distributions with one variate are defined as:

$$g(x|\mu, \sigma) = \frac{1}{\sqrt{2\pi\sigma^2}} e^{\frac{-(x-\mu)^2}{2\sigma^2}} \tag{3.8}$$

where μ and σ are the mean and variance (standard deviation) of the normal distribution. Foreground and background can be estimated with the help of:

$$\left|\mu_{t+1} - X_{t+1}\right| < k\sigma_{t+1} \tag{3.9}$$

If the above conditions are met, the pixel is considered a background pixel; otherwise the pixel is considered a foreground pixel. In video sequences with continuous gray levels and infrequent changes in illumination, the single Gaussian method could be used to represent bimodal backgrounds. In indoor surveillance, where the illumination is almost constant, and in outdoor surveillance, where the amount of light and the background clutter change greatly, it is a useful technology. Every frame updates the mean and variance. Such challenges and constraints can be addressed using multivariate Gaussian distributions.

- *Multivariate Gaussian Distribution*

Mixtures of Gaussians and Gaussian mixture model are both terms used to describe multivariate Gaussian distributions. A Gaussian mixture is used to model each pixel in the background model. In general, the model is parametrized by the mean, variance, and probability of each Gaussian component. Figure 3.3 shows the multivariate Gaussian distribution. A complex model is created by combining all the simple models.

$$\left(X\middle|\mu, \sum\right) = \sum_{k=1}^{K} \omega_k \aleph\left(X\middle|\mu_k, \sum_k\right) \tag{3.10}$$

where $\aleph(X|\mu_k, \sum_k)$ is referred to as the Gaussian distribution components and ω_k is referred to as the mixing coefficient.

Multivariate Gaussian distribution easily adopts the dynamic scenes and handles gradual and sudden changes in illuminations. Further it can be expressed as:

$$\aleph\left(X\middle|\mu_k, \sum_k\right) = \frac{1}{\sqrt{2\pi \sum}} e\left\{\frac{-1}{2}(X-\mu)^T \sum\nolimits^{-1}(X-\mu)\right\} \tag{3.11}$$

where $\aleph(X|\mu, \sum)$ represent the normal distribution. μ and Σ are the mean and covariance of the multivariate Gaussian distribution. Figure 3.4 shows the flowchart of the proposed object detection and tracking algorithm. The algorithm is based on the mixture of Gaussian. The proposed intrinsic improvements (modification made in the initialization, the maintenance, and the foreground detection step) of the Gaussian model make it robust to handle not only clutter and dynamic backgrounds but also efficiently handle partial occlusions and shadows. Another improvement in terms of extrinsic improvements (external approaches/algorithms/tools to perform the results) can handle the dataset noise and outlier of the video sequences. An extrinsic improvement significantly improves the performance evaluation of the proposed algorithm.

3.4.1 Frame Analysis

The proposed algorithm shows that every video sequence is to be converted into frames, and the algorithm requires initial known or unknown number of frames to develop the background model. If the initial frames of the video are clean or free from the moving object, it requires very few frames to develop background model. If the moving object is present in initial frames, it requires some subsequent frames to develop the background model. Our proposed algorithm uses probabilistic statistical Gaussian mixture model to estimate the background model for foreground detection.

3.4.2 Preprocessing

Preprocessing has always been an important phase for the entire process. Preprocessing not only improves the performance evaluation, but it also helps to reduce the execution time. If we could use monochrome images rather than colored image, it saves lots of processing time as color needs three color channels instead of one 8-bit channel. In some video sequences like camera jitter, pan-tilt-zoom (PTZ),

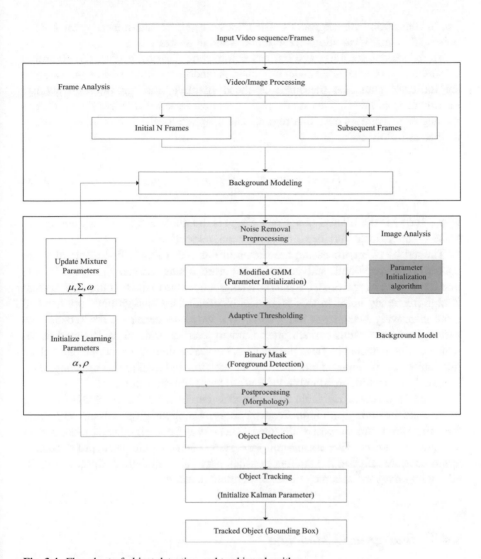

Fig. 3.4 Flowchart of object detection and tracking algorithm

and low frame rate, a change in illumination creates noisy images. To process all those sequences, image smoothness is required to reduce image noise and enhance the detection and tracking accuracies.

The detection and tracking complexities can be increased by background motions such as leaves floating on water, twinkling and glinting of water, etc. By processing it through a filtering environment, it is possible to reduce the resolution to handle such challenges. All the high-frequency components are suppressed by a low-pass pass filter, which provides smooth images of moving objects. In addition to decreasing the scattering noise, low resolution also reduces the generation of false positives due to nonstationary backgrounds. The filter type will be determined by the type of noise and intensity variations (both indoors and outdoors).

To reduce dataset noise and outliers before re-processing begins, the algorithm uses the adaptive local noise reduction filter. A random variable's mean and variance are the basic statistical measures. A mean adaptive filter provides an average measurement of intensity across the region, while a variance adaptive filter indicates the degree of contrast over that region. The adaptive local noise reduction filter is defined as follows:

$$\widehat{f}(x, y) = g(x, y) - \frac{\sigma_n^2}{\sigma_L^2}[g(x - y) - \mu_L] \tag{3.12}$$

where $\widehat{f}(x, y)$ is the filtered image and $g(x, y)$ is the noisy image. σ_n^2, σ_L^2, and μ_L are noise variance, local variance, and local mean, respectively.

The results of preprocessing are shown in Fig. 3.5. CDnet 2014 standard video dataset is compared both with and without preprocessing. As shown in Fig. 3.5, the first row has suffered from the dynamic background. Due to the floating leaves, there is high-frequency noise in this sequence. Nonstationary backgrounds are handled more effectively because of the results. The video sequence on the second row suffers from continuous camera jitter. Camera jittering leads to background noise, which can be detected as false positives. A low-pass filtering technique can reduce jitter noise. It will improve the accuracy of detection and tracking by reducing high-frequency noise with an adaptive local noise reduction algorithm.

In the third row, a dynamic background is depicted by the twinkling of water surface and gleaming from high-frequency noise. Using the preprocessing approach, the foreground can be easily detected. There will be a significant reduction in background noise. The results of this study demonstrate that preprocessing approaches are efficient in handling nonstationary data and dataset noise. Additionally, it improves the accuracy of the foreground detection.

3.4.3   Background Modeling

The Gaussian mixture model is used to generate the background model. It consists of initializing the background model parameters, maintaining the background model, and detecting the foreground model. Our proposed algorithm has certain intrinsic improvements, such as initializing the model parameters using the parameter optimization algorithm (i.e., ω, μ, σ, and ρ). Using such an algorithm, parameters are generated for all sequences, and the algorithm can be used both indoors and outdoors. Detection of foreground objects has also been improved. For detection of the foreground objects, adaptive thresholding achieves better segmentation, which leads to robust foreground detection rather than predefined thresholding.

- *Mixture of Gaussians*

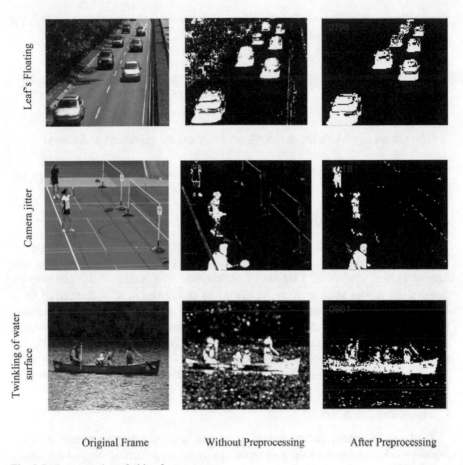

Leaf's Floating   Camera jitter   Twinkling of water surface

Original Frame   Without Preprocessing   After Preprocessing

Fig. 3.5 Preprocessing of video frame

The Gaussian mixture model is parameterized by the mean vectors μ_i, covariance matrices Σ_i, and mixture weights ω_i; the initial mixture model is represented by:

$$\lambda = \left\{ \mu_i, \sum_i, \omega_i \right\} \tag{3.13}$$

Each pixel of the background is characterized by RGB color space and models each pixel $\{X_1, X_2, \ldots, X_t\}$ as a mixture of K Gaussian distributions. The probability of the current pixel value is expressed by multimodalities as follows:

$$P(X_t) = \sum_{i=1} \omega_{i,t} \times N\left(X_t, \mu_{i,t}, \sum_{i,t}\right) \tag{3.14}$$

where K is the number of Gaussian distributions, $\omega_{i,t}$ is the weight of the ith Gaussian at time t, $\mu_{i,t}$ is the mean of the ith Gaussian at time t, $\Sigma_{i,t}$ is the covariance matrix of the ith Gaussian at time t, t is referred to as the frame index or time, and N is the Gaussian probability density function.

The mixture weight $\omega_{i,t}$ satisfies the constraints as follows:

$$\sum_{i=1}^{K} \omega_{i,t} = 1 \tag{3.15}$$

The Gaussian probability density function N can be expressed as follows:

$$N\left(X_t, \mu, \sum\right) = \frac{1}{(2\pi)^{n/2}|\sum|^{1/2}} e^{-\frac{1}{2}(X_t-\mu)^T \sum^{-1}(X_t-\mu)} \tag{3.16}$$

- *Number of Components of K Gaussian Distributions*

The multimodality of the backgrounds is determined by K. The number of components is also determined by memory and computing power, so it is also known as the number of Gaussian distributions. The value of the components decides the model such as $K = 1$ referred to as unimodal, $K = 2$ referred to as bimodal, and $K = 3$–5 considered as multimodal. Usually, the number of distributions K is predefined. Various approaches like estimation with the help of online algorithm, stochastic approximation procedure, dynamic variations, ISODATA algorithm, etc. are available for the determination of the number of distributions K. For the proposed algorithm $K = 3$ (multimodal backgrounds) remains fixed over the entire detection process.

- *Parameter Initialization*

Background model parameters ω, μ, σ and the learning parameters α, ρ are initialized with the help of the parameter optimization algorithm. Figure 3.6 shows the detailed flowchart of the parameter optimization algorithm. For indoor and outdoor surveillance system, it is proposed in the literature that the parameters can be initialized with the help of various approaches such as k-means algorithm, expectation maximization (EM) algorithm, background reconstruction algorithm, etc. For every video sequence, this proposed algorithm estimated the model parameters, and every time the background model is initialized with the estimated parameters and hence the proposed background model is to handle clutter and dynamic background. The flowchart indicates that every time the initialization depends on the number of iterations and the ground truth image. The initialization is achieved with the help of the $f_{\text{minsearch}}$ and finds minimum of unconstrained multivariable function using derivative-free method. After the initialization of the entire set of parameters, load the mixture parameter file using the available parameters. For both the indoor and outdoor environments, the available parameters are optimized. 50 iterations give optimum values of mixture parameters and time, while if the iterations are increased, the algorithm takes more time with no significant difference in parameter values and its iterations are reduced and the parameter values change significantly; considering

Fig. 3.6 Flowchart of
parameter initialization

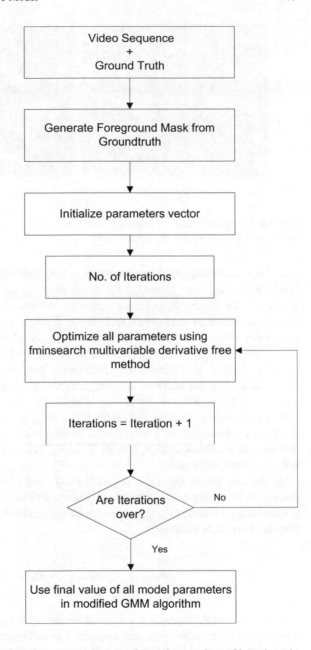

the tradeoff between time and optimum parameter values, the number of iterations is
set to 50.

- *Foreground Detection*

Foreground detection is the process of determining foreground-background
pixels. The process of segmenting a scene into foregrounds and backgrounds is

(a)   (b)

Fig. 3.7 Foreground mask for a scene (area of motions is presented as the color in foreground mask). (**a**) Input image; (**b**) Foreground mask

also known as foreground segmentation. Using the initialized background model, foreground masks are produced by subtracting the current frame from the background model. There are generally stationary elements that serve as the background, like roads, buildings, and plants, and they will stay stationary over the course of the entire detection and tracking process. It simply alters the pixel values of the stationary background pixels under the influence of illumination variations and weather changes. The camera may mistakenly perceive leaves floating, twinkling, and gleaming water surfaces as foreground objects in some cases. Foreground objects are pixels that are moving or likely to be moving in a scene. If a moving foreground object remains stationary for a long period of time, it can also be considered a background. Foreground detection is also known as motion detection. A foreground-motion mask is illustrated in Fig. 3.7. Motion is represented by mask colors in the foreground.

If the distribution shows more pixel weight and poor or weak variance, then it belongs to the background rather than to be considered as the foreground. Therefore, the first B Gaussian distributions are considered as a background as it exceeds certain threshold limits as follows:

$$B_{\text{back}} = \arg\min{}_b \left(\sum_{i=1}^{b} \omega_{i,t} > T \right) \tag{3.17}$$

T is a measure of the minimum portion of the data that should be accounted for by the background. If we select the small value of T, it will be considered as a single modal. If the value of T is very high, then it will be considered as a multimodal. The value of T is highly sensitive to the scene environment, especially for the outdoor scene; it ties to the learning rate α. If the distribution does not satisfy, the above relation is to be considered as a foreground distribution. For every new frame at time $t + 1$, match test is performed for every pixel. A pixel matches a Gaussian distribution if it satisfies the Mahalanobis distance:

$$\sqrt{\left(\left(X_{t+1} - \mu_{i,t}\right)^T \cdot \sum_{i,t}^{-1}\left(X_{t+1} - \mu_{i,t}\right)\right)} < k\sigma_{i,t} \qquad (3.18)$$

where k is the constant threshold equal to 2.5 and X_{t+1} is a color feature of pixel of a new frame at time $t + 1$.

After match test is found with one of the K Gaussian distributions, the distribution is identified as a background, and if it fails to match it, the distribution is considered as foreground distribution.

- *Foreground Detection Using Adaptive Threshold*

The process of segmenting pixels into foreground and background is called foreground segmentation. In the literature, several segmentation methods are available, such as intensity-based segmentation, region-based segmentation, texture-based segmentation, edge-based segmentation, motion-based segmentation, etc. In all approaches, intensity-based threshold provides quick and simple classification. Static threshold can lead to some difficulties, for example, if it is too large it will detect objects that are not present in the ground truth, resulting in false positives and decreasing accuracy. When the threshold value is too high, the small differences in color may not be detected. On the other hand, if the threshold level is too high, the system will not track the moving objects from the ground truth, it will instead produce false negatives, and therefore there will be a decrease in recall. When the foreground color is like the background distribution, a low threshold won't be able to detect it, as well as causing permanent noise. By using adaptive thresholds, rather than predefined thresholds, all constraints will be considered, as well as the performance evaluation. The adaptive threshold is expressed as follows:

$$T_a = \min[A] \qquad (3.19)$$

where $A = (\mathrm{Fg}_\omega \times \mathrm{Fg}_v) + (\mathrm{Bg}_\omega \times \mathrm{Bg}_v)$, T_a is referred to as the adaptive threshold, Fg_ω and Bg_ω are referred to as the foreground and background weights, and Fg_v and Bg_v are referred to as the foreground and background variance. The foreground and background weights are calculated as:

$$\mathrm{Fg}_\omega = \frac{\sum_{i=1}^n H(1,i)}{\sum_{i=1}^n H} \qquad (3.20)$$

$$\mathrm{Bg}_\omega = \frac{\sum_{i=1}^g H(i,255)}{\sum_{i=1}^n H} \qquad (3.21)$$

where H = image (frame) histogram, g = gray level, and n = number of pixels.

The foreground variance and the mean μ are calculated as:

$$\mathrm{Fg}_v = \frac{\sum_{i=1}^{n}(A_i - \mu)^2 H(1, i)}{\sum_{i=1}^{n} H} \tag{3.22}$$

where $\mu = \frac{\sum_{i=1}^{n} H(1, i) \times A}{\sum_{i=1}^{n} H}$ and $A_i \in (1, i)$.

The background variance and the mean μ are calculated as:

$$\mathrm{Bg}_v = \frac{\sum_{i=1}^{g}(A_i - \mu)^2 H(i, 255)}{\sum_{i=1}^{n} H} \tag{3.23}$$

where $\mu = \frac{\sum_{i=1}^{n} H(i, 255) \times A}{\sum_{i=1}^{n} H}$ and $A_i \in (i, 255)$.

Our main contribution as far as the intrinsic improvements of the model are concerned is the development of adaptive threshold techniques and a better method for selecting the threshold value to distinguish background objects from foreground objects. It can be seen from Fig. 3.8 that the first sequence has been affected by clutter and dynamic background elements, which have inadvertently been rendered as floating leaves and baselines. With adaptive thresholds, moving vehicles can be segmented efficiently as compared to ground truth. It suffered from low frame rates in the second sequence. With a low frame rate, it would be very difficult to detect motion segmentation for video sequences as scenes change very quickly in every frame and it would be extremely difficult to generate a background. A comparison of segmented and ground truth results indicates significant improvements. Despite having a dynamic background, the third sequence is still capable of detecting moving objects, even when faced with dynamically changing and high-intensity backgrounds. The background model for every sequence is developed and updated continuously with each new pixel and frame.

- *Background Model Maintenance/Upgradation*

A line IIR upgradation algorithm is used to update the background model to adopt the changes in the scene as the process continues. This is a learning process at every pixel and frame level, and it must be done online. By using the previous background, foreground binary mask, and current frame, the maintenance component can update the background. Adopting to the changes that occurred in the background is the process, but it isn't exclusive to backgrounds. During periods of prolonged stationary foregrounds, model parameters may need to be updated. Various background maintenance schemes, such as blind, selective, and fuzzy adaptive, are discussed in literature. In parameter maintenance schemes, choosing the right learning rate and the appropriate maintenance mechanism and determining the frequency of model updates are the key issues. When light changes and dynamic changes in the video sequence occur, a highly reliable and good background maintenance model is required.

After successful detection of moving objects, background model parameters need to be updated for every incoming pixel and frame. Using the matched test equation

Fig. 3.8 Results for foreground detection using adaptive threshold. (**a**) Original frame. (**b**) Background. (**c**) Ground truth. (**d**) Foreground mask

(3.18), the pixel is classified as a background or foreground, and then based on classification, updation is required. If the match is found with one of the K Gaussian distributions, then the pixel belonging to background and the model parameters weight, mean, and the variance need to be updated. The new background model is estimated from the initial model and represented as follows:

$$p(X|\bar{\lambda}) \geq p(X|\lambda) \qquad (3.24)$$

where λ is referred to as the initial model, X is the new color frame, and $\bar{\lambda}$ is referred to as the new updated background model.

- *Learning Rate α and ρ*

A constant learning rate α and ρ is used to update the model parameters. α is used to update the mixture weights, while ρ is used to update the mean and variance. The parameter optimization algorithm is used to estimate the learning rates. A weighted value updated rate may also be used to refer to the learning parameters. With the learning parameters, scenes can be adapted to changes more quickly and additional roles can be added to handle scene dynamics. Selections of the learning parameters are as follows: fixed or dynamic, statistical, and fuzzy.

It must be precisely selected for the fixed parameters as it is used throughout all sequences. For different Gaussian distributions, different statistical learning rates must be used. By using such an approach, performance will be improved. During the foreground detection, fuzzy membership values are obtained for each pixel, and finally, the learning rate is calculated. The selection of the learning rate depends on the application, and certain applications require the algorithm to adapt fast enough to the changing scene. In a few applications, the algorithm needs to store the temporal history of pixels, and the slow learning rate is the only way to accomplish this. To adapt to the changes as well as preserve the important temporal history of the pixels, our algorithm assigns different learning rates for every sequence.

3.5   Pixel Belongs to the Background

Once the match test is found, the pixel is classified as the background, so it needs to update background model parameter such as weight ω, mean μ, and variance σ. Updated weight using learning rate α and mixture weight is computed as follows:

$$\omega_{i,t+1} - (1 - \alpha)\omega_{i,t} + \alpha \qquad (3.25)$$

where α is the learning rate and it is very sensitive to scene dynamics and illuminations and its value is closely associated with the threshold, and $\omega_{i,t+1}$ is referred to as the updated mixture weight. Updated mean and variance using learning rate ρ and the mean and the variance are updated as follows:

$$\mu_{i,t+1} = (1 - \rho)\mu_{i,t} + \rho \cdot X_{t+1} \tag{3.26}$$

where $\mu_{i,\,t\,+\,1}$ is referred to as the updated mean.

$$\sigma^2_{i,t+1} = (1 - \rho)\sigma^2_{i,t} + \rho \cdot (X_{t+1} - \mu_{i,t+1}) \cdot (X_{t+1} - \mu_{i,t+1})^T \tag{3.27}$$

where $\sigma^2_{i,t+1}$ is referred to as the updated variance and $X_{t\,+\,1}$ is a color feature of pixel of new frame at time $t + 1$. The learning rate is calculated as follows and its lower value will make the convergence slow. It is used to update current distributions and reduces computation power and provides faster Gaussian tracking.

$$\rho = \alpha \cdot \eta \left(X_{t+1}, \mu_i, \sum_i \right) \tag{3.28}$$

3.6 Pixel Belongs to the Foreground

If the match test fails, then the pixel is classified as the foreground. For foreground pixel it is necessary to upgrade the mixture weight, while the mean and variance of the background model remains unchanged. Updated weight using learning rate α and mixture weight under the foreground classification is calculated are follows:

$$\omega_{j,t+1} = (1 - \alpha)\omega_{j,t} \tag{3.29}$$

where $\omega_{j,t+1}$ is referred to as the updated mixture weight.

In Fig. 3.9, the significance of the learning rate is explained. In all three cases, the learning parameters not only maintain the significance of their temporal history but also handle dynamic scenes in a dynamic manner. The result shows that the learning parameters deal efficiently with low and high lighting, and the sudden variations of illumination caused by changes in illumination are also improved. With our proposed approach, detection and tracking are more accurate than the ground truth as well as providing faster adaptation. Due to the parameter initialization algorithm, the learning rate can be adapted for varying environments.

3.6.1 Preprocessing

The noise in foreground objects must be reduced once the foreground object has been detected. Extrinsic improvements focus only on improving the model's performance and results. By removing the smallest segment among multiple segmented regions and merging object regions, the largest regions are removed. Using

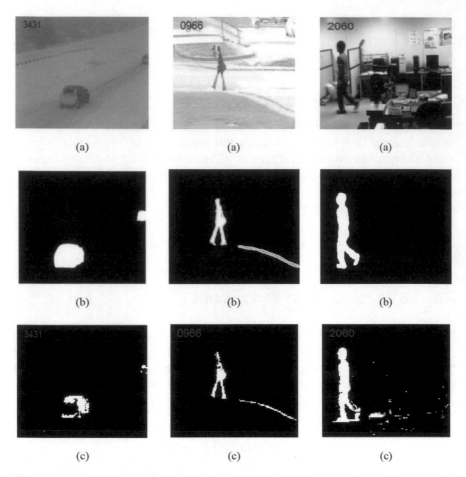

Fig. 3.9 Significance of learning rate—adopting scene changes. (**a**) Original frame. (**b**) Ground truth. (**c**) Effect of learning rate for frame with low illumination, high illumination, and sudden change

morphological closing, the foreground noise is reduced by postprocessing. During morphology operations, tiny gaps located within moving objects are filled and residual noise is reduced. The morphological closing is dilation followed by erosion. The background pixels that are touching an object pixel are changed into object pixels.

Figure 3.10 illustrates the effects of postprocessing after foreground detection. As a result, postprocessing eliminates isolated regions and fills in small gaps. Postprocessing improves the result significantly over the ground truth. The processes of morphological closing-dilation and erosion can improve the results because background pixels are replaced by pixels of the foreground associated with the foreground, and erosion removes isolated foregrounds. In addition to reducing false positives, the results show that the postprocessing also enhances the system

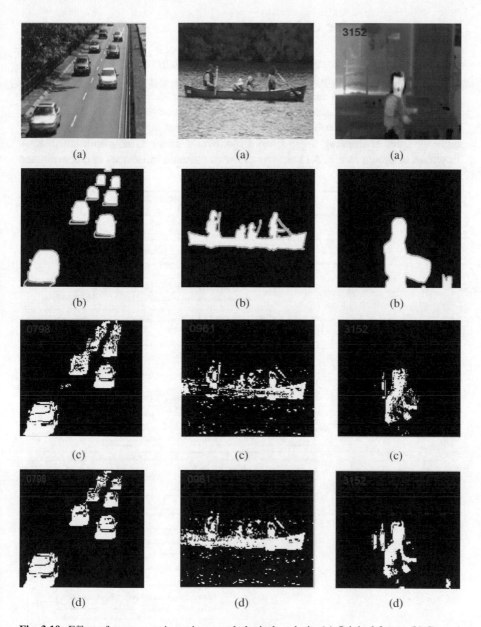

Fig. 3.10 Effect of postprocessing using morphological analysis. (**a**) Original frame. (**b**) Ground truth. (**c**) Without postprocessing. (**d**) With postprocessing

precision by improving accuracy when it comes to detecting and tracking false objects.

3.7 3D Monocular Object Detection

An essential part of using the monocular is estimating 3D data. Monocular images require 3D reconstruction by interpreting the pixels as surfaces with depth and orientation. There is no need for perfect depth estimates when texture mapping combined with a rough sense of geometry provides convincing detail. Monocular images can be used for robotic navigation systems and autonomous driving assistance systems. Estimating depth from a monocular image is an important task. A flowchart is shown in Fig. 3.11 showing how to estimate the shapes and coordinates

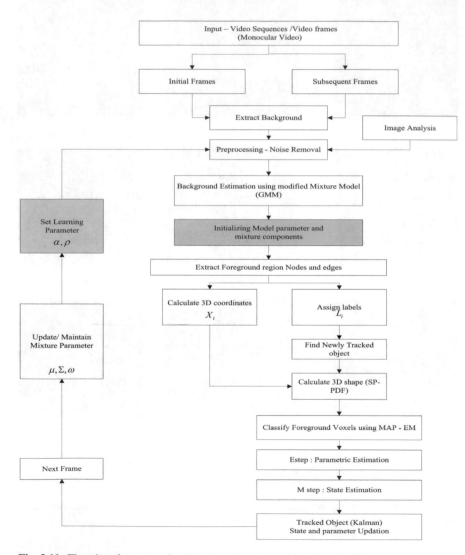

Fig. 3.11 Flowchart for monocular 3D object detection and tracking algorithm

based on the monocular image. There are various ways to estimate 3D information [3–32], such as RGBD cameras, stereoscopic 3D, CAD models, simple conversions from 2D to 3D, and monocular 3D. Monocular images can easily be used to estimate depth. These monocular cues include variations in texture, gradients in texture, occlusion, known object sizes, haze, etc.

Over the past few years, 3D object detection and tracking methods have improved dramatically. Each approach relies on three factors: depth sensors, feature points, and texture. Since 2D information does not include altitude information, 3D detection is more robust than 2D. Researchers often use a geometry- and appearance-based approach to analyze data. In this presented algorithm it estimates 3D coordinates and position in monocular images with a geometry-based approach. Researchers who utilize monocular multiview can also detect 3D objects with occlusion. It is easy to identify objects without altitude since the multiview estimates depth cues strongly. The presented algorithm assumes certain assumptions when it comes to absolute detection in monocular images, such as:

- Track objects from one camera.
- Track objects offline.
- Moving object distances can be taken up to a certain level.
- Objects of small silhouettes are to be ignored.

3.7.1   Foreground Voxel Classification

In 3D object detection and tracking in monocular sequence, it is mandatory to classify foreground and background objects in the same way as 2D detection. 3D pixel is referred to as voxel and it can be expressed as $X = (x, y, z)$ denoting the Cartesian coordinates of a scene. The classifications of foreground voxels for the moving objects in monocular sequence are considered as the assumptions of mutual distributions. The mutual distributions are expressed as $p(X, l)$, where X is referred to as 3D coordinate of a foreground voxel and $l \in \Gamma$. Γ is defined as the foreground region and their voxels are represented as the set of labels $\Gamma = \{1, 2, \ldots, m\}$. m is referred to as the number of objects. The joint or mutual and the conditional distributions can be shown as follows:

$$p(X, l) = p(l)p(X|l) \tag{3.30}$$

where $p(X|l)$ is referred to as the likelihood function and $p(l)$ is referred to as the prior. The prior is used as a mixing coefficient in the expectation maximization framework. The prior of the expectation maximization algorithm and one of the important model parameters is expressed as $p(l) = \pi_l$, and it must satisfy the following constraints:

$$\sum_{l \in \Gamma} \pi_l = 1^0 \leq \pi_l \leq 1, \forall l \in \Gamma \tag{3.31}$$

The 3D position of the moving object at time t has been considered and it has been expressed as ψ_l^t. The conditional probability of the above Eq. (3.30) is considered as the conditional likelihood function, and it is defined as the probability of posterior distribution and X given ψ_l^t and is given by:

$$p(X|l) \sim p(X|\psi_l^t) \tag{3.32}$$

where $p(X|\psi_l^t)$ is referred to as the semiparametric probability distribution function (PDF) whose parameters are ψ_l^t and it represents the 3D silhouette of the object. If the silhouette of l remains unchanged from previous time $t-1$ to t, voxel X belongs to l and it represents the voxels belonging to dense foreground.

For the maximum a posteriori-expectation maximization (MAP-EM), define prior for the ψ_l^t as the multivariate Gaussian distributions:

$$p(\psi_l^t) \sim N\left(\psi_l^t \middle| \psi_l^{t-1}, \sum\right) \tag{3.33}$$

where ψ_l^{t-1} is referred to as the position of the object as t is near $t-1$ and Σ is referred to as the covariance matrix of normal distribution.

3.7.2 Maximum a Posteriori-Expectation Maximization

Expectation maximization (EM) is an iterative process to calculate the maximum likelihood estimation of a given parameter. Usually expectation maximization (EM) is used to estimate the Gaussian mixture model (GMM). Sometimes it has also been used to learn the optimal mixture of fixed models. Based on probability distributions, MAP-EM is used to classify the complete set of foreground voxels. For the complete set of 3D coordinates of the foreground, voxels are represented as χ. The main objective is to calculate the maximum a posteriori resolution for the presented modified Gaussian mixture models with the help of the defined probability distribution functions. The initial model parameters are expressed as follows:

$$\Theta = \{\pi_l, \psi_l^t\}, l \in \Gamma \tag{3.34}$$

Using expectation maximization and the iteratively derived updated model parameters Θ^{t+1} from the current approximation Θ^t, the following are the EM steps for 3D model estimation:

- *E Step*:

In estimation step, the posterior distribution for the Θ^t is calculated and can be expressed as:

$$\Theta^t = p(l|X, \Theta^t) \tag{3.35}$$

where the entire foreground voxels belong to the set of 3D coordinates ($\forall X \in \chi$).

• *M Step*:

In the maximization, calculate the Θ^{t+1} using the sum of the log likelihood Q function and the logarithm of prior:

$$\Theta^{t+1} = \max_{\Theta} R(\Theta|\Theta^t) \tag{3.36}$$

where $R(\Theta|\Theta^t)$ can be further calculated as follows:

$$R(\Theta|\Theta^t) = Q(\Theta|\Theta^t) + \ln p(\Theta) \tag{3.37}$$

where $Q(\Theta|\Theta^t)$ is referred to as the Q function and $\ln p(\Theta)$ is referred to as the logarithm of prior. The Q function can be estimated from the following:

$$Q(\Theta|\Theta^t) = \sum_{l \in \Gamma} \ln p(\chi, l|\Theta)p(l|\chi, \Theta^t) \tag{3.38}$$

The above equation can be further simplified by the Bayes' theorem as follows:

$$\ln p(\chi, l|\Theta) = \sum_{X \in \chi} \{\ln p(l) + \ln p(X|l)\} \tag{3.39}$$

$$p(\chi, l|\Theta) = \frac{p(l)p(\chi|l, \Theta)}{p(\chi)} \tag{3.40}$$

The approximate $\ln p(\Theta)$ using the prior $\ln p(\psi_l^t)$ gives the following:

$$\ln p(\Theta) \sim \sum_{l \in \Gamma} (\psi_l^t) \tag{3.41}$$

Finally, the 3D model parameters are updated as follows:

The prior can be updated as: $\pi_l^t \to \pi_{lu}^t$.
An object's 3D position can be updated as: $\psi_l^t \to \psi_{lu}^t$.

3.8   Results and Discussion

By using a statistical background subtraction method (modified Gaussian mixture model), motion segmentation and 2D object detection can be accomplished. The presented approach can be used both indoors and outdoors and can handle various constraints. The results of the 2D detection are shown in Figs. 3.12 and 3.13. A comparison of the results for different outdoor video sequences can be seen in Fig. 3.12. To evaluate the performance of the proposed algorithm, it is being tested over standard video datasets like ViSOR, CDnet 2014, and PETS 2009. A clutter background, high illumination, pedestrians partially obscured, and near- and far-field objects adversely affect all the sequences in Fig. 3.12. The proposed detection method accurately detects all objects in almost all video sequences. False positives and false negatives can be reduced effectively using statistical background models.

The detection results for indoor video sequences are shown in Fig. 3.13. The presented algorithm is tested using several challenging indoor video datasets such as ViSOR, CDnet 2014, CAVIAR, and PETS 2006. Figure 3.13 comprises sequences that suffer from partial occlusion, intermittent motion, high indoor reflection, high illumination, and low contrast thermal imaging. The proposed detection method accurately detects all objects in almost all video sequences. The statistical background modeling effectively reduces false positives and false negatives. The results indicate that both environments can be handled efficiently by the proposed algorithm and that it can also handle a range of challenges. Detection accuracy was improved by a significant margin.

3.9   Summary of Chapter

By enhancing both intrinsically and extrinsically the background Gaussian mixture model, a modified approach is presented. The combination of indoor and outdoor environments is handled by applying optimized parameter techniques to initialize the mixture parameters. In this chapter, we discuss methods for detecting foreground objects in 2D and 3D monocular scenes. This chapter presents several objectives aimed at improving foreground detection:

- Adaptive thresholding ensures robust foreground detection even when clutter and dynamic backgrounds are present.
- By reducing dataset noise and outliers and improving performance evaluation parameters, the adaptive local noise reduction filter reduces the dataset noise and outliers.
- In motion segmentation, postprocessing plays a crucial role. Morphological closing improves segmentation accuracy.
- With probability density functions and MAP-EMs, foreground objects are estimated in 3D by estimating the voxels in monocular sequences.

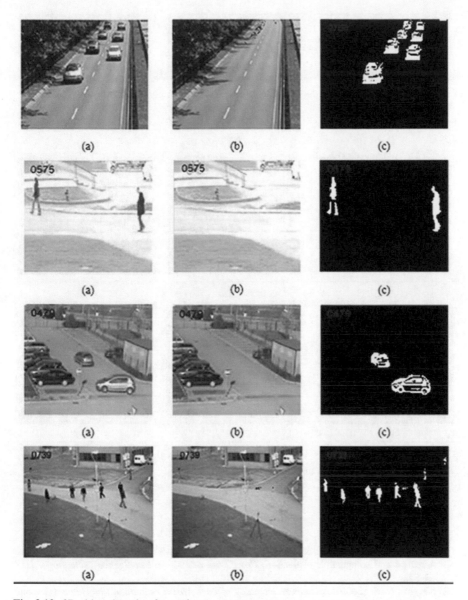

Fig. 3.12 2D object detection for outdoor sequences

A mixture model based on semiparametric component distributions is used for 3D shapes for both indoor and outdoor environments in the proposed 2D and 3D object detection algorithms. According to the performance evaluation, there is a significant improvement over the ground truth. Various improvements in the Gaussian mixture model, along with the learning parameters, allow the algorithm to cope with clutter, dynamic, and changing light conditions effectively.

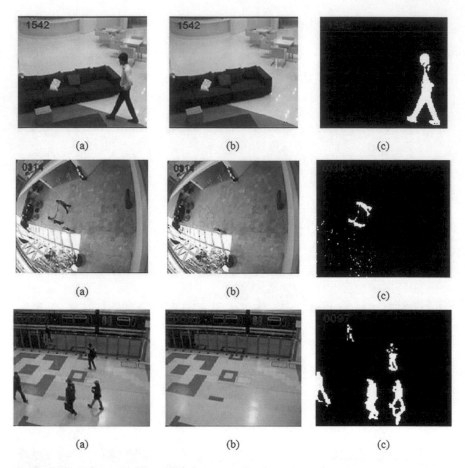

Fig. 3.13 2D object detection for indoor sequences

References

1. Xu, Y., Dong, J., Zhang, B., & Xu, D. (2016). Background modeling methods in video analysis: A review and comparative evaluation. *CAAI Transactions on Intelligence Technology, 1*(1), 43–60.
2. Naraghi, N. S. (2009). *A comparative study of background estimation algorithms* (Doctoral dissertation). Eastern Mediterranean University.
3. Stauffer, C., & Grimson, W. E. L. (2000). Learning patterns of activity using real-time tracking. *IEEE Transactions on Pattern Analysis and Machine Intelligence, 22*(8), 747–757.
4. Carr, P., Sheikh, Y., & Matthews, I. (2012, October). Monocular object detection using 3D geometric primitives. In *European conference on computer vision* (pp. 864–878). Springer.
5. Sekii, T. (2016). Robust, real-time 3D tracking of multiple objects with similar appearances. In *Proceedings of the IEEE conference on computer vision and pattern recognition* (pp. 4275–4283). IEEE.

6. Najafi, H., Genc, Y., & Navab, N. (2006, January). Fusion of 3D and appearance models for fast object detection and pose estimation. In *Asian conference on computer vision* (pp. 415–426). Springer.
7. Crivellaro, A., Rad, M., Verdie, Y., Moo Yi, K., Fua, P., & Lepetit, V. (2015). A novel representation of parts for accurate 3D object detection and tracking in monocular images. In *Proceedings of the IEEE international conference on computer vision* (pp. 4391–4399). IEEE.
8. Xiang, Y., Song, C., Mottaghi, R., & Savarese, S. (2014, September). Monocular multiview object tracking with 3D aspect parts. In *European conference on computer vision* (pp. 220–235). Springer.
9. Brachmann, E., Krull, A., Michel, F., Gumhold, S., Shotton, J., & Rother, C. (2014, September). Learning 6D object pose estimation using 3D object coordinates. In *European conference on computer vision* (pp. 536–551). Springer.
10. Chliveros, G., Pateraki, M., & Trahanias, P. (2013, July). Robust multi-hypothesis 3D object pose tracking. In *International conference on computer vision systems* (pp. 234–243). Springer.
11. Tejani, A., Tang, D., Kouskouridas, R., & Kim, T. K. (2014, September). Latent-class Hough forests for 3D object detection and pose estimation. In *European conference on computer vision* (pp. 462–477). Springer.
12. Wojek, C., Roth, S., Schindler, K., & Schiele, B. (2010, September). Monocular 3D scene modeling and inference: Understanding multi-object traffic scenes. In *European conference on computer vision* (pp. 467–481). Springer.
13. Ess, A., Müller, T., Grabner, H., & Van Gool, L. (2009, September). Segmentation-based urban traffic scene understanding. In *BMVC* (Vol. 1, p. 2). The British Machine Vision Association and Society for Pattern Recognition.
14. Brostow, G. J., Shotton, J., Fauqueur, J., & Cipolla, R. (2008, October). Segmentation and recognition using structure from motion point clouds. In *European conference on computer vision* (pp. 44–57). Springer.
15. Tu, Z., Chen, X., Yuille, A. L., & Zhu, S. C. (2005). Image parsing: Unifying segmentation, detection, and recognition. *International Journal of Computer Vision, 63*(2), 113–140.
16. Hoiem, D., Efros, A. A., & Hebert, M. (2008). Putting objects in perspective. *International Journal of Computer Vision, 80*(1), 3–15.
17. Byeon, M., Oh, S., Kim, K., Yoo, H. J., & Choi, J. Y. (2015). Efficient spatio-temporal data association using multidimensional assignment in multi-camera multi-target tracking. In *BMVC* (pp. 1–68). The British Machine Vision Association and Society for Pattern Recognition.
18. Iwashita, Y., Kurazume, R., Hasegawa, T., & Hara, K. (2006, May). Robust motion capture system against target occlusion using fast level set method. In *Proceedings 2006 IEEE international conference on robotics and automation, 2006. ICRA 2006* (pp. 168–174). IEEE.
19. Luo, X., Berendsen, B., Tan, R. T., & Veltkamp, R. C. (2010, August). Human pose estimation for multiple persons based on volume reconstruction. In *2010 20th international conference on pattern recognition* (pp. 3591–3594). IEEE.
20. Chen, X., Kundu, K., Zhang, Z., Ma, H., Fidler, S., & Urtasun, R. (2016). Monocular 3D object detection for autonomous driving. In *Proceedings of the IEEE conference on computer vision and pattern recognition* (pp. 2147–2156). IEEE.
21. Chen, X., Kundu, K., Zhu, Y., Berneshawi, A. G., Ma, H., Fidler, S., & Urtasun, R. (2015). 3D object proposals for accurate object class detection. In *Advances in neural information processing systems* (pp. 424–432). IEEE.
22. Karpathy, A., Miller, S., & Fei-Fei, L. (2013, May). Object discovery in 3D scenes via shape analysis. In *2013 IEEE international conference on robotics and automation* (pp. 2088–2095). IEEE.
23. Krahenbuhl, P., & Koltun, V. (2015). Learning to propose objects. In *Proceedings of the IEEE conference on computer vision and pattern recognition* (pp. 1574–1582). IEEE.
24. Lee, T., Fidler, S., & Dickinson, S. (2015). A learning framework for generating region proposals with mid-level cues. In *International conference on computer vision* (Vol. 2). IEEE.

25. Ohn-Bar, E., & Trivedi, M. M. (2015). Learning to detect vehicles by clustering appearance patterns. *IEEE Transactions on Intelligent Transportation Systems, 16*(5), 2511–2521.
26. Pepik, B., Stark, M., Gehler, P., & Schiele, B. (2015). Multi-view and 3D deformable part models. *IEEE Transactions on Pattern Analysis and Machine Intelligence, 37*(11), 2232–2245.
27. Xiang, Y., Choi, W., Lin, Y., & Savarese, S. (2015). Data-driven 3D voxel patterns for object category recognition. In *Proceedings of the IEEE conference on computer vision and pattern recognition* (pp. 1903–1911). IEEE.
28. Leibe, B., Cornelis, N., Cornelis, K., & Van Gool, L. (2007, June). Dynamic 3D scene analysis from a moving vehicle. In *2007 IEEE conference on computer vision and pattern recognition* (pp. 1–8). IEEE.
29. Vacchetti, L., Lepetit, V., & Fua, P. (2004). Stable real-time 3D tracking using online and offline information. *IEEE Transactions on Pattern Analysis and Machine Intelligence, 26*(10), 1385–1391.
30. Kelly, P. H., Katkere, A., Kuramura, D. Y., Moezzi, S., & Chatterjee, S. (1995, January). An architecture for multiple perspective interactive video. In *Proceedings of the third ACM international conference on multimedia* (pp. 201–212). ACM.
31. Sato, K., Maeda, T., Kato, H., & Inokuchi, S. (1994, February). CAD-based object tracking with distributed monocular camera for security monitoring. In *Proceedings of 1994 IEEE 2nd CAD-based vision workshop* (pp. 291–297). IEEE.
32. Jain, R., & Wakimoto, K. (1995, May). Multiple perspective interactive video. In *Proceedings of the international conference on multimedia computing and systems* (pp. 202–211). IEEE.

Chapter 4
Object Tracking

4.1 Introduction

To track an object, successive pictures are taken, which allow for analysis of its behavior and estimation of its track. The task of tracking objects is one of the most challenging in machine learning and computer vision. Various approaches are presented in the literature for tracking objects. In an object tracking algorithm, the major goal is to analyze automated driving assistance and robot navigation, and the major motivation is the availability of high-end computers with inexpensive, high-quality video cameras. Each video analysis aims to identify the moving objects, track them from frame to frame, and monitor their movement. There are also applications of tracking such as recognition of motion-based objects, video indexing, multimedia, automated surveillance, traffic monitoring, vehicle navigation, counting of vehicles and people, augmented reality, and gesture recognition, among others.

Object tracking involves tracking the entire visible foreground or segmented objects, describing the trajectory of each tracked object. Generally, object tracking methods are classified according to object representation and features of the image. In addition, there are numerous approaches to tracking objects, which can be classified into four major groups: contour tracking, feature tracking, model tracking, and region tracking. There are three main tracking approaches: point tracking, silhouette tracking, and kernel tracking. It is also possible to categorize deterministic and probabilistic approaches to object tracking. Based on the deterministic approach, the frame/image is compared to the modeled object and all possible regions are identified, following which the object is selected. A deterministic approach is mean shift filtering. As a tracking method, the probabilistic approach uses a state-space model. Probabilistic approaches are exemplified by particle filtering. In other words, tracking is the process of determining the trajectory of a foreground object in a frame or image.

© The Author(s), under exclusive license to Springer Nature Switzerland AG 2022
N. Ghedia et al., *Moving Objects Detection Using Machine Learning*, SpringerBriefs
in Electrical and Computer Engineering,
https://doi.org/10.1007/978-3-030-90910-9_4

Fig. 4.1 Different object
tracking approaches

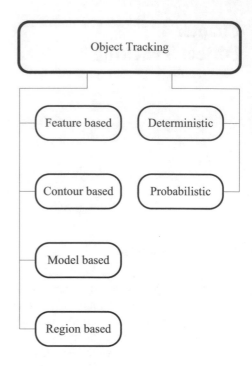

Fig. 4.1 Different object tracking approaches

A general classification of tracking approaches is shown in Fig. 4.1. Generally, most tracking methods belong to these categories as stated in the literature. There are three types of methods for tracking, namely, point tracking, kernel tracking, and silhouette tracking. A two-step approach to tracking objects is followed. A model for the moving target must be defined, and a prediction about the moving objects in the current scene must be derived from knowledge about the moving objects in the previous scene. This technique is recursively applied to narrow down the error and update the tracking model in each scene. In order to get motivated in motion tracking, researchers make certain assumptions, such as object motion is flat and without abrupt changes, objects can be tracked at a constant speed, and the algorithm is aware of the number of objects, appearance, size, and silhouette of the objects. Various tracking algorithms are subject to the same constraints or challenges.

- The unrestrained movement of either static or moving cameras would make tracking the object difficult because it would lead to issues regarding size, silhouette, and intensity. Such dynamic variations require an appropriate model.
- Inter-object occlusion, as well as occlusion caused by stationary or nonstationary backgrounds, including buildings, trees, bridges, etc., creates serious tracking issues.
- It would be more difficult to track areas with limited camera views or distributed locations.
- The object tracking process is also affected by errors in the detection phase.

- It would also be more complicated to track nonrigid and uttered objects due to noise.
- It's also challenging to track real-time information.

Single or multiple objects can be tracked using object tracking algorithms. In general, point-based Kalman filtering and particle filtering provide moderate to high tracking accuracy while having a relatively low computational cost. A multi-object occlusion can be handled using the said approach. Kernel-based mean shift and support vector learning are also able to track multiple objects, but their accuracy will be lower. Kernel-based approaches handle partial occlusions efficiently. Shape matching based on silhouettes provides excellent accuracy at the expense of computational time. It can track multiple objects in partial and full occlusion environments and can handle multiple objects.

4.2 Kalman Filtering

The Kalman filter is an excellent tool for estimating the predicted values. Kalman filters are optimal estimators. Based on tentative, indirect, and erroneous observations, the parameters of concern are recursively estimated. When it comes to word filtering, it sometimes creates confusion. It's not intended to suppress the data measurements, but to find the best estimate from the noisy incoming data. Therefore, the filter is used for filtering out the noise in the data and predicting the best estimate. Its critical features include excellent performance due to optimization, the ability to manage and formulate filtering in real time, and the ease with which it is implemented. This filter is also referred to as the best linear estimator.

Kalman filter is a mathematical process that uses a series of equations and input data. The interactive process quickly estimates an object's position, velocity, and other information. A measurement contains an unforeseen error, random error, uncertainty, or variation. A bunch of data inputs will help us estimate the real value quickly. By analyzing the variations and uncertainty of the data inputs, the Kalman filter takes a few of the inputs from the bunch of data and narrows them down to their true values. There is no truth to the data coming in, only a close approximation of it. The Kalman filter is also known as a sequential state estimation. Kalman filters have the following important applications:

- Use of radar to track targets, i.e., missiles and aircrafts
- Global positioning system (GPS) for satellite tracking, calculating the position and velocity of the satellite, and unmanned vehicle tracking
- Navigation, i.e., moving objects and autonomous driving assistance
- Various computer vision applications, i.e., video surveillance, depth and velocity measurements, feature and cluster tracking, and image filtering
- Fault-tolerant mechanism and data fusion
- Smoothing and phase-locked loop (PLL) for space vehicle
- Robot localization

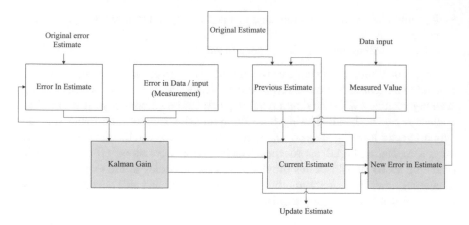

Fig. 4.2 Simplified block diagram of Kalman filter

The generalized block diagram of the Kalman filter is shown in Fig. 4.2. Kalman gain, current estimation, and new estimation are the three main components of the Kalman filter. The Kalman gain can be calculated using the estimation error and measurement error. It is used for calculating both new error estimates and current estimations. Error estimates depend on estimating the original errors. With the Kalman gain, the previous estimation, and the measured value, the current state can be estimated. The measured value is strongly influenced by the bulk data inputs. To calculate the new error in the estimate, the current estimate and Kalman gain are used. The output of each iteration of the current estimate will be based on the updated estimate. Using the Kalman filter, a new error in estimate is calculated each time, so the recursive algorithm will limit the error and estimate the true value quickly. The Kalman filter estimates true value iteratively by three different calculations:

- Kalman gain
- Current estimate
- New error in estimate

4.2.1   Kalman Gain

A Kalman gain is one of the important parameters of the Kalman filter that helps us understand it better. To update the estimate, new measurements are needed. Calculating the Kalman gain involves dividing the error in the estimate by the sum, i.e., the error in the estimate plus E_{est} the error in the measurement E_{mea}. Kalman gains are usually between 0 and 1.

$$K_g = \frac{E_{est}}{E_{est} + E_{mea}} \tag{4.1}$$

where K_g is referred to as the Kalman gain, E_{est} is referred to as the error in estimate, and E_{mea} is referred to as the error in measurement. The gain would be between $0 \leq K_g \leq 1$.

- *Significance of Kalman Gain*

The Kalman gain implies that the measurement is fairly accurate, and the estimates become unstable, or there is a large amount of uncertainty. This can be corrected by reducing the Kalman gain while maintaining the measurement's accuracy. For large Kalman gains, a high-accuracy measurement or very limited error in the measurement results in a large contribution to the update given by the measurement.

4.2.2   Current/New Estimate

A current estimate is defined as equality to the previous estimate plus the Kalman gain times the difference between the measured value and the previous estimate.

$$est_t = est_{t-1} + K(mea - est_{t-1}) \tag{4.2}$$

where est_t and est_{t-1} are referred to as the current and previous estimate. K_g is referred to as the Kalman gain and mea is referred to as measurement.

4.2.3   New Error in Estimate

The error in the estimate is calculated as the previous error in the estimate multiplied by the error measurement and divided by the sum of the error measurement and the previous error in estimate.

$$E_{est(t)} = \frac{E_{mea} \cdot E_{est(t-1)}}{E_{mea} + E_{est(t-1)}} \tag{4.3}$$

where $E_{est(t)}$ and $E_{est(t-1)}$ are referred to as the error in the current and previous estimate and E_{mea} is referred to as the error in the measurement.

The error in the estimate is alternatively calculated as:

$$E_{\text{est}(t)} = \left(1 - K_{\text{g}}\right) \cdot E_{\text{est}(t-1)} \tag{4.4}$$

where K_{g} is referred to as the Kalman gain.

4.2.4   Kalman Filter Process Derivations-State-Space Derivation

The ultimate solution to discrete data linear filtering was provided by R.E. Kalman [1] in 1960. Using previous measurements and observations, this is a mathematical process that consists of iteratively recursively estimating the process. Based on the linear stochastic difference equation [2], the Kalman filter model estimates the state of a discrete time-controlled process:

$$x_k = A \cdot x_{k-1} + B \cdot u_{k-1} + w_k \tag{4.5}$$

and the measurement equation is defined as:

$$z_k = H \cdot x_k + v_k \tag{4.6}$$

Table 4.1 shows the explanation of the symbols used in the above equations.

A prior state estimate of the state vector at time step k is denoted as $\widehat{x}_k^- \in \mathfrak{R}^n$ and a posteriori state estimate of the state vector at step k given measurement z_k is denoted as $\widehat{x}_k \in \mathfrak{R}^n$. A prior and a posteriori error can be defined as follows:

Table 4.1 Symbol explanations used in Kalman filter

Parameter	Description	Properties
x_k	State vector—important model parameters such as position, velocity, etc.	$x \in \mathfrak{R}^n$
A	State transition matrix—relates to the state at the previous time step $k - 1$ with the current time step k	$A \in \mathfrak{R}^{n \times n}$
B	Optional control inputs—transforms control vector to state vector (transforms force into acceleration)	$B \in \mathfrak{R}^{n \times 1}$
u_k	Control input vector (force, throttle settings)	$u \in \mathfrak{R}^l$
w_k	It is a random variable that defines the process noise. It is assumed to be independent and has a normal probability distribution with covariance matrix Q	$p(w) \sim N(0, Q)$
z_k	Measurement vector	$z \in \mathfrak{R}^m$
H	State transformation matrix, maps into measurement domain	$H \in \mathfrak{R}^{m \times n}$
v_k	It is a random variable that defines the measurement noise. It assumed to be independent and has a normal probability distribution with covariance matrix R	$p(v) \sim N(0, R)$

$$\bar{e}_k = x_k - \widehat{x}_k^{\,-} \tag{4.7}$$

$$e_k = x_k - \widehat{x}_k \tag{4.8}$$

A prior estimate error covariance matrix is calculated as follows:

$$P_k^- = E\left[\bar{e}_k \bar{e}_k^{\,T}\right] \tag{4.9}$$

A posteriori estimate error covariance matrix is calculated as follows:

$$P_k = E\left[e_k e_k^{\,T}\right] \tag{4.10}$$

The Kalman filter computes a posteriori state estimate $\widehat{x}_k^{\,-}$ and a linear combination of prior estimate $\widehat{x}_k$, and a weighted difference between an actual measurement z_k and a measurement prediction $H\widehat{x}_k^{\,-}$ is shown as follows:

$$\widehat{x}_k = \widehat{x}_k^{\,-} + K(z_k - H\widehat{x}_k^{\,-}) \tag{4.11}$$

where K is referred to as the Kalman gain and the difference $(z_k - H\widehat{x}_k^{\,-})$ is referred to as innovation or residual.

The above Eq. (4.11) is known as the current or new estimate for the filter process. The Kalman gain is also referred to as a blending factor and it is used to minimize the a posteriori error covariance. The Kalman gain is calculated as follows:

$$K_g = \frac{P_k^- H^T}{H P_k^- H^T + R} \tag{4.12}$$

where $P_k^- H^T$ is referred to as the error in the estimate and R is referred to as the error in the measurement.

The Kalman filter is used to estimate the linear system where the state is assumed to be Gaussian. Figure 4.3 shows the recursive approach for the Kalman filter. The process shows how the iterative process will quickly estimate the true value by narrowing down the errors. The Kalman gain, the current estimation, and the new error in estimation equations are the three major equations of the Kalman state-space derivations. The error in the estimate can be calculated as follows:

$$P_k = \frac{P_k^- \cdot R}{P_k^- + R} \tag{4.13}$$

where P_k^- is referred to as the previous error in the estimate and R is referred to as the error in the measurement.

The error in the estimate can also be calculated as follows:

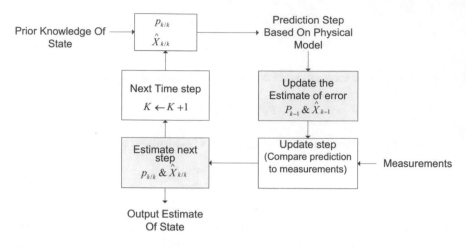

Fig. 4.3 Recursive approach of Kalman filter

$$P_k = \left(1 - K_g H\right) P_k^{-} \tag{4.14}$$

Kalman filters are used only for linear equations in state space. It is not capable of handling distributions other than Gaussian. When tracking nonlinear equations, particle filters are the most effective. The extended Kalman filter solves all the problems that are associated with Kalman filtering as well as solving nonlinear stochastic difference equations. An occluded object is calculated by using a Kalman filter.

Using the Kalman filter, an estimate is estimated at some time, and the feedback is then derived from (noisy) measurements. The complete operation of the Kalman filter can be categorized into two stages:

- *Time update*: The time update uses the current state and the error covariance estimates to determine the a priori estimates for the next/ahead of time step operation. This stage can also be called a predictor or a prediction stage.
- *Measurement update*: The measurement update used a prior estimate to compute the posterior or new error in estimation. It is also referred to as the corrector or correction stage.

Figure 4.4 shows the prediction and correction stage for the Kalman filter. It also shows the iterative and recursive linear approach to minimize the means of squared error.

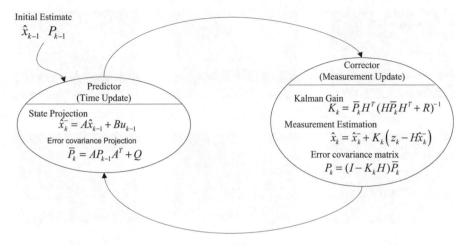

Fig. 4.4 State diagram for prediction and correction stage of the Kalman filter

4.3 Object Tracking Using Kalman Filtering

By using this approach, motion segmentation can be achieved with the help of statistical background subtraction, and tracking can then be done with the help of Kalman filters that recursively enclose the segmented foreground. The approach can be applied to both indoor and outdoor environments, and it can accommodate a variety of constraints. Tracking results are shown in Figs. 4.5, 4.6, 4.7, and 4.8. The tracking results for the various indoor video sequences are shown in Fig. 4.5. It is being tested on standard video datasets such as ViSOR, CDnet 2014, and CAVIAR to evaluate the performance of the proposed algorithm.

The first sequence was hampered by the varying lighting conditions, the second sequence was under constant illumination but suffered from static occlusion, and the third sequence was heavily hampered by the clutter background and heavy reflections. This tracking approach correctly detects and tracks nearly all moving objects in all video sequences. False positives and false negatives can be reduced effectively using statistical background models.

The tracking results for outdoor video sequences are shown in Fig. 4.6. Various challenging outdoor video datasets are used to test this algorithm. A background video sequence with outdoor clutter is presented as the first sequence, a dynamic background video sequence with continuous leaf floating and weaving is presented as the second sequence, and the third and final video sequence has a very low contrast and low frame rate. The proposed tracker is capable of tracking slow-moving and fast-moving objects as well as detecting partially occluded moving objects in each sequence. Based on the results, the proposed algorithm is effective at handling the various challenges and improves detection and tracking accuracy.

The tracking results for near-, mid-, and far-field moving objects are shown in Fig. 4.7. A series of outdoor video sequences ViSOR was used to evaluate the

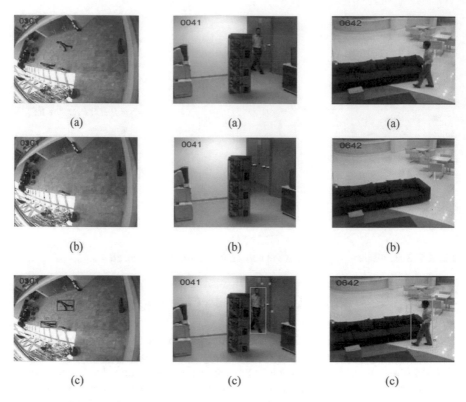

Fig. 4.5 Object tracking in indoor environment. (**a**) Original frame. (**b**) Best background. (**c**) Tracked object

performance of this algorithm. Objects in the near and mid field are clearly tracked in the first sequence. The second sequence demonstrates the tracker's ability to efficiently track far-field objects. The tracking results demonstrate that this algorithm tracks objects in dynamic and outdoor clutter environments efficiently. Moreover, it can handle both stationary and nonstationary backgrounds.

The results of the monocular 3D object tracking are shown in Fig. 4.8. The algorithm is tested on standard indoor and outdoor video datasets like PETS 2009, APIDIS, and ViSOR. Pedestrians are clearly detected in the first sequence. It is the second sequence of the video dataset where this algorithm has the greatest challenge due to the shiny, reflective finish. The third video sequence again shows the ability to recognize moving objects under varying illumination patterns. Results indicate that the proposed algorithm can handle a variety of constraints for monocular 3D object tracking.

Fig. 4.6 Object tracking in outdoor environment. (**a**) Original frame. (**b**) Best background. (**c**) Tracked object

4.4 Summary of Chapter

The Kalman filter is used in this chapter to track linear objects. A recursive analytical tracking approach using the discrete Kalman filter is presented along with the various tracking strategies. It provides the most accurate tracking among all methods. Through previous and current measurements, the spatial dependence of physical parameters is estimated. Kalman filters find the correct estimate for a linear system under the influence of white Gaussian noise and emphasize the following features:

- Determine the future location of the moving target.
- Reduce noise introduced by inaccuracy or uncertainty.
- Provide trajectories to single or multiple objects.

The tracking approach has been evaluated both indoors and outdoors using standard video datasets. Various challenges were handled successfully and tracking accuracy improved. Based on performance evaluations, it has been shown that this

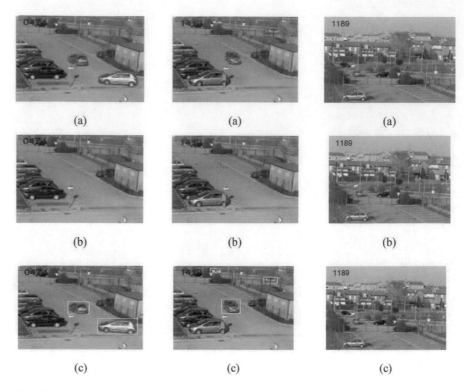

Fig. 4.7 Near-field, mid-field, and far-field object tracking. (**a**) Original frame. (**b**) Best background. (**c**) Tracked object

algorithm is effective at tracking moving objects in the near field, mid field, and far field.

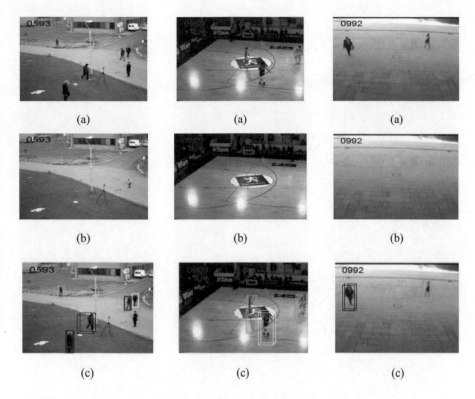

Fig. 4.8 3D object tracking. (**a**) Original frame. (**b**) Best background. (**c**) Tracked object

References

1. Kalman, R. E. (1960). A new approach to linear filtering and prediction problems. *Journal of Basic Engineering, 82*(1), 35–45.
2. Bishop, G., & Welch, G. (2001). An introduction to the Kalman filter. *Proceedings of SIGGRAPH, Course, 8*(27599-23175), 41.

Chapter 5
Summary of the Book

5.1 Introduction

A two-dimensional and monocular three-dimensional object detection and tracking algorithm has been developed and examined in this book for both indoor and outdoor surveillance. The presented algorithm is evaluated with standardized datasets to track single and multiple moving objects, and its performance is compared to other similar methods. Two major modules in the presented algorithm are (a) object detection in three dimensions and two dimensions and (b) object tracking by means of discrete Kalman filters. For every sequence of videos, the background model is estimated using the improvised Gaussian mixture model. To optimize the mixture and learning parameters for GMM, the parameter optimization algorithm is used. Under various constraints and challenging conditions, adaptive thresholding is used to segment the foreground objects following parameter optimization. Once the dataset noises, outliers, and other noises have been removed, the algorithm performs better using both the preprocessing (adaptive local noise reduction filter) and postprocessing (morphological closing) methods. Discrete Kalman filtering has been used to track the detected objects. In both indoor and outdoor environments, the presented tracking method can track two-dimensional and monocular three-dimensional objects.

5.2 Outcome of the Presented Work

The presented system performs well in all the aspects examined in the book. A two-dimensional and monocular three-dimensional object detection and tracking algorithm is presented and discussed and its results are presented. It discusses the

© The Author(s), under exclusive license to Springer Nature Switzerland AG 2022
N. Ghedia et al., *Moving Objects Detection Using Machine Learning*, SpringerBriefs
in Electrical and Computer Engineering,
https://doi.org/10.1007/978-3-030-90910-9_5

intrinsic and extrinsic factors that improve the Gaussian mixture model. The major findings of this study are as follows:

- The presented algorithm detects objects in indoor and outdoor environments with complex and dynamic backgrounds efficiently and with high reliability.
- The algorithm presented is capable of handling partial occlusions and a certain number of shadows.
- Optimized mixture parameters (intrinsic improvement) reduce background clutter and nonstationary elements.
- Monocular 3D works with objects having similar appearances.
- The adaptive threshold (intrinsic improvement) can detect foreground objects.
- Motion detection accuracy significantly increases when noise, outliers, and false positives are reduced through pre- and postprocessing (extrinsic improvements).
- This algorithm can cope with any object class and can also cope with a wide range of lighting conditions.
- The presented algorithms manage crowded scenes efficiently and are invariant to camera perspectives.
- At the same time, the presented algorithm is capable of handling multiple objects at the same time.

The presented algorithm has some limitations like difficulty detecting shadows in video frames with small texture variations in foreground and background, inability to detect stationary foreground objects, and difficulty tracking fully occluded objects. Foreground detection performance drops off significantly in extreme and rapid light changes, and detecting edges of objects in very low-resolution video is difficult.

5.3   Future Research Direction

In this study, a wide range of object classes with other dataset characteristics are targeted by object detection and tracking algorithm. However, there are still several areas in need of further improvement to make the algorithm more generic. The following are some potential future directions:

- Intrinsic improvement in Gaussian mixture model—maintain the mixture parameters by means of dynamic approach.
- Different values for learning parameters and other approaches to calculate the learning parameters make it easy to manage the sudden illumination variations.
- The algorithm can be more robust by adding other pixel features such as texture and edges.
- For real-time surveillance application, DSP or other hardware realization is required.

Index

© The Author(s), under exclusive license to Springer Nature Switzerland AG 2022
N. Ghedia et al., *Moving Objects Detection Using Machine Learning*, SpringerBriefs
in Electrical and Computer Engineering,
https://doi.org/10.1007/978-3-030-90910-9

Printed in the United States
by Baker & Taylor Publisher Services